The Engineering of
Human Joint
Replacements

The Engineering of Human Joint Replacements

J.A. McGeough
The University of Edinburgh, UK

WILEY

This edition first published 2013
© 2013 John Wiley & Sons, Ltd

Registered office
John Wiley & Sons Ltd, The Atrium, Southern Gate, Chichester, West Sussex, PO19 8SQ, United Kingdom

For details of our global editorial offices, for customer services and for information about how to apply for permission to reuse the copyright material in this book please see our website at www.wiley.com.

Library of Congress Cataloging-in-Publication Data

McGeough, J. A. (Joseph A.), 1940- author.
The engineering of human joint replacements / by J.A. McGeough. – 1
    1 online resource.
    Includes bibliographical references and index.
    Description based on print version record and CIP data provided by publisher; resource not viewed.
    ISBN 978-1-118-53684-1 (ePub) – ISBN 978-1-118-53685-8 – ISBN 978-1-118-53690-2 (MobiPocket) – ISBN 978-0-470-74027-9 (hardback)   1. Artificial joints.   2. Arthroplasty. 3. Biomedical engineering.   I. Title.
    RD686
    617.5'80592–dc23
2013022090

A catalogue record for this book is available from the British Library.

ISBN: 978-0-470-74027-9

Set in 10/12.5pt Palatino by Aptara Inc., New Delhi, India.

Printed and bound in Malaysia by Vivar Printing Sdn Bhd

1   2013

Contents

Preface

I wish to thank Mr James Christie, Consultant Orthopaedic Surgeon at the Royal Infirmary of Edinburgh (RIE), who first suggested to me that I could contribute engineering solutions to clinical problems.

Our encounter happened by chance. Following an injury to my knee some weeks earlier, I was given an appointment to see him. Until we met I had been researching mainly in manufacturing processes. During a visit to a major aircraft engine manufacture to discuss cooperation in electrochemical machining (ECM) (a method of shaping tough heat-resistant alloy metals that are difficult to cut by established mechanical techniques), I stumbled on a stairway and to stop myself falling I straightened my right leg, and suffered a sharp pain in my knee. As the pain continued for some weeks I eventually saw my doctor who referred me to the RIE; that is how I met Mr Christie.

He solved the problem with my painful knee – the abrupt jerk had reactivated an old sports injury to my cartilage and associated wear and tear over the years. After asking me what I did for a living, and when I explained my professional interests, he brought me an X-ray of an intramedullary nail, the design for which he felt needed changing. Over the next few months, with a colleague, James Holmes, I designed and had manufactured a prototype in titanium alloy, which was the material I had been discussing with the aircraft engine manufacturer when my injury had occurred. The prototype was put into the hands of a medical device company for evaluation. My cooperation with orthopaedic surgeons had begun.

Other collaborative projects followed with Mr (now Professor) Charles Court-Brown and Dr Richard McCalden. We investigated the effects of age on the mechanical properties of cortical and cancellous human bone. With Professor Dugald Gardner, and later Mr John Keating, there began a long-standing cooperation on engineering studies of the meniscus.

Following a long stint of the headship of the Department of Mechanical Engineering at the University of Edinburgh, in 1991 I was granted a sabbatical period of six months. It was arranged by Mr Christie that I could be based in the academic unit of the Orthopaedic Surgery Department in the Edinburgh Royal Infirmary. In addition to meeting other surgeons in their daily work, including Miss (now Professor) Margaret McQueen, it provided me with the privilege of attending operations in order to understand the role that engineering can have in orthopaedic surgery.

On return to full-time university duties, the collaborative programmes continued, firstly in the Princess Margaret Rose Orthopaedic Surgery Hospital and then in the new Royal Infirmary of Edinburgh. While the Crichton Laboratory, devoted to orthopaedic engineering, had been opened in the Department of Mechanical Engineering in 1989, it had become increasingly evident that close geographical proximity of engineering research students and assistants to clinicians is essential for research work of this nature to flourish. To that end, with cooperation from academic colleagues in orthopaedic surgery, we were fortunate to be able to establish the then-named 'Edinburgh Orthopaedic Engineering Centre' (EOEC) within the College of Medicine and Veterinary Medicine's Chancellor's Building, with its own experimental and office facilities.

This book is an account of engineering's place in that part of orthopaedic surgery that deals with human joint replacement. It is based on impressions gathered in consultations with orthopaedic surgeons over many years, observations made during the attendance at surgical operations described above, and the insight provided from innumerable discussions with research clinicians and engineering researchers, both post-graduate and undergraduate, whom I have introduced to this field through their projects for PhD, MPhil, MSc, BEng and MEng degrees.

The book is aimed primarily at the last-mentioned group. They receive little, if any, formal introduction to the field in their teaching courses, which are so heavily constrained by accreditation requirements. Yet I have never known one who has not risen enthusiastically to the task of investigating some aspect of engineering relevant to orthopaedic surgery. This enthusiasm has constantly been encouraged by their meetings with clinicians, who have always made time to see them.

The book is therefore not an account of the latest research findings in this field. Instead it lays the foundations from which research can arise, although in the course of its writing some new relevant developments have been noted and introduced. For some of these I am grateful to the referees who vetted the outline plan at the outset. For others, I have drawn on the series of event proceedings organized by the Engineering in Medicine and Health division of the Institution of Mechanical Engineers.

The subject is clearly cross-disciplinary. Before the work proper began, the question of a co-author from orthopaedic surgery was raised with the publisher and clinicians. The latter are busy people in their daily duties and the added burden of writing such a tome with the time needed would have added too much to their already heavy workload. Another engineer as a co-author might have helped, but consultation revealed that this would have been possible in specialist areas and not across the entire spectrum of its contents. Instead the book has been written by me, as one author. It is therefore one engineer's perspective on the subject. I hope I can be excused if I have not given sufficient scope to topics that might be of direct interest to specialists in areas discussed in the book. Nonetheless I am grateful to colleagues who have read each of its draft chapters and made many suggestions for improvement. They include Mr C. Howie, Professors D.L. Gardner and S. Hinduja, Dr E. Keane, Drs J. Atkinson, R. Heinemann and G. McGuinness, and Messrs A. Room and M. Wright.

The staff of the library of the Royal College of Surgeons of Edinburgh, especially Mrs Marianne Smith and Mr Steven Kerr, have been most helpful in arranging access for me in my researches for the book. Special thanks are due to Mr Colin Howie, senior consultant orthopedic surgeon, who arranged for me to attend operations at the RIE in order to deepen my understanding of joint-replacement surgery.

The book has been typed by Ms Diane Reid. Miss Jiayi Shu prepared the diagrams, with useful contributions from L. Delimata, C. Fraser and C. Macmillan. Ms L. Delimata provided sterling support during the final stages of preparation of the book. I am grateful to authors, and others acknowledged in the text for permission to reproduce photographs and figures.

The staff at Wiley Engineering Publishing, notably Ms Debbie Cox, the late Nicky Skinner, Ms Liz Wingett, and Mr Tom Carter provided professional and patient encouragement from the outset to completion, for which I am much appreciative. Within the School of Engineering at the University of Edinburgh, Professors Alan Murray and Ian Underwood are thanked for their support.

Books demand time for writing. In this task patience is needed by those closest to the author. My wife Brenda possesses this virtue, for which I am continually grateful. The other members of my family, Andrew and Karen McGeough, Elizabeth and Barry Keane with Patrick and Thomas, and Simon and Louise McGeough with William and Amelia, remain my steadfast supporters, always interested in what I do.

J.A. McGeough
Edinburgh

1

Introduction

Movements of the human body are controlled by its skeletal structure, which is made up of bones, joints, muscles, tendons and nerves. Even in healthy bodies, these parts of the skeletal structure can develop disorders, the symptoms of which can be pain, stiffness, swellings, deformity, loss of function and changes in sensitivity. Of these symptoms, pain is the most common.

The outcome of the sensation of pain can be impaired movement of the body; yet reduced movement could also be due to stiffness, localized to a particular joint, or more generally at more than one joint. A common form of stiffness is 'locking', the inability to complete a movement of the body. This condition can arise from mechanical effects, such as torn or damaged parts. For example, in the knee, vigorous athletic activity can cause a tear in the 'meniscus' (cartilage), causing both pain and locking.

Variations from normality ('deformity') in the skeletal system can take forms such as wide hips, short limbs, round shoulders and curvature of the spine; their magnitude can change with time. When joints do not function properly, muscle weakness and joint instability can be a consequence.

Damage to bones and joints, and their associated nerve supplies, can cause a change in sensitivity, which may present itself as numbness, tingling or pain. A common example is a collapsed intervertebral disc that causes pressure on a neighbouring part of the skeletal structure, leading to back ache; another is a trapped nerve in the skeleton, which usually causes pain, and loss or reduction in function or movement of the body. A frequent cause of pain stems from inflammation of a joint caused by arthritis; swelling and stiffness may come from injury, infection or degeneration due to wear and tear. When arthritis is the cause, key joints in the body, notably the knee, become unable to

The Engineering of Human Joint Replacements, First Edition. J.A. McGeough.
© 2013 John Wiley & Sons, Ltd. Published 2013 by John Wiley & Sons, Ltd.

support the upper body weight, and everyday physical movements become impaired.

Osteoarthritis (OA) is one of the main types of arthritis causing these difficulties. This disease leads to progressive degeneration in the cartilage of the knee, lessening the ability of the cartilage to cushion the joint from impact and provide articulation of the knee. The rate of advancement of OA is influenced by body weight, the extent of physical activity, age and genetic and orthopaedic abnormalities. When pain, stiffness and knee movement reach unbearable levels, joint replacement has to be considered. The incidence and prevalence of OA increases with age and it is also associated with obesity. The World Health Organisation (WHO) defines obesity in terms of body mass index (BMI), calculated by dividing weight (kg) by square of height (m^2). A BMI of respectively 25–30, 30–40, and more than 40 kg/m^2 categorizes the conditions of overweight, obese, and morbidly obese. Changulani et al. (2008) quote UK government statistics that between the ages of 55 and 64, about 20% of men and 33% of women are obese. Studies by Busija et al. (2007) indicate that pain due to OA is more prevalent in adults who are obese or overweight. The meniscus (cartilaginous tissue) acts as the shock absorber of the knee joint, bearing 50 to 70% of load on the knee. Ordinary movements such as walking downstairs entail forces four times that of the body weight being imposed on the meniscus. When the knee joint has to bear additional bodyweight due to obesity, meniscal degradation can arise. This will be explored more fully in Chapter 3.

Other joints not directly involved in load bearing can also be affected by obesity. The accumulation of body fat may promote inflammatory effects in tissue, causing joints to deteriorate.

The World Health Organisation (2003) reports that the combination of energy-dense foods and physical inactivity have contributed to a threefold increase in obesity since 1980, in countries ranging from North America to Australia and China. In the United States, more than 70% of people over 60 years of age are considered to be obese. By 2015, 700 million people world-wide are expected to be obese, according to the BBC (2008).

Obese people with a BMI of more than 30 kg/m^2 are encouraged to lose weight. A reduction in weight by 5 kg in obese people has been claimed to reduce by 24% their need for surgery associated with OA of the knee (Williams and Fruhbeck 2009). Others have suggested that a BMI greater than 40 kg/m^2, or morbid obesity, might contraindicate surgery (Horan 2006).

People who undergo knee replacement surgery are less likely to have a less successful outcome if they have a BMI of more than 40 kg/m^2, compared to those with a BMI below 30 kg/m^2 (Amin et al. 2006).

Total knee replacement is more common in women than men, and this has led to these implants being specifically adapted to fit the geometrical features of the female knee.

Rheumatoid arthritis (RA) is another common cause of joint replacement. It affects about 1 in 100 people at some stage, usually between the ages of 40 and 60, and the condition is three times more likely to occur in women than in men. It is more common in smokers and in those who consume high quantities of red meat and caffeine.

In RA, the body's immune system dysfunctions. The inflammation that ensues causes damage to tissue, bone and neighbouring ligaments, thinning out the cartilage and adversely affecting the synovial fluid that serves to lubricate the joint spaces. The joints become swollen, painful and deformed. They can no longer perform their proper function.

Every joint in the body can be affected by degenerative conditions like OA and RA. The need for joint replacements can also be caused by trauma, such as car accidents, falls, or by athletic injury, causing broken bones or fractures, or other damage to the joint structure. A fracture or break in a bone can disrupt the supply of blood to bone, which adversely affects its healing. This can lead to 'avascular necrosis' or 'osteonecrosis', that is, death of the bone. In the hip, this complication is another leading cause of joint replacement and is most often observed in patients between the ages of 30 and 50.

The lifespan of a primary hip prosthesis is estimated to be 10 to 15 years (although new designs are extending this period), after which revision may be required. The need for revision can stem from many causes including weakening of the original femur bone owing to age or disease, or 'aseptic' loosening of the prosthesis within the parent bone. The effects of age and gender on primary and revision hip replacement surgery are presented in Table 1.1. See also the National Health Service (NHS) National Services Scotland (2012). The worldwide need for hip joint replacement is evident from Table 1.2.

Many attempts have been made to replace human joints with manmade substitutes. In the 1860s, knee replacements were undertaken aimed at restoring the normal functions of this joint. A platinum and rubber replacement for the shoulder joint was produced in 1893. In the early twentieth-century, work on hip replacements began, the head of the femur being replaced by ivory and then later acrylic, the prosthesis being fitted with a stem that was positioned in the femoral neck. These early devices proved of limited efficacy and were duly abandoned, although some, albeit limited, progress continued to be made. Nonetheless an increasing insight was gained into the requirements for effective joint replacement.

Table 1.1   Effects of age and gender on number (N) of primary and revision hip replacements per 100 000 of population.

		Men (primary)	Women (primary)	Men (revision)	Woman (revision)
Age group (years)	*Number (N)*	*N*	*N*	*N*	*N*
< 35	167	208	41	53	
35–39	164	168	29	52	
40–44	257	259	46	91	
45–49	412	434	74	97	
50–54	857	1075	155	203	
55–59	1355	1690	247	283	
60–64	2241	2895	420	474	
65–69	2969	3815	590	662	
70–74	2479	4466	589	931	
75–79	1987	4566	583	1192	
80–84	1114	2735	423	977	
85+	519	1593	300	995	
Total	14 521	23 904	3497	6010	

Source: National Health Service (NHS) (2000) *Information Centre for Health and Social Care*, Office for National Statistics, London.

Table 1.2   Number of hip replacements in Europe, North America and Australasia.

Country	Total	Primary	Revision	Revision (%)
Sweden	15 679	13 942	1737	11
Denmark	8292	7244	1048	13
Germany	218 173	196 391	21 782	10
Italy	64 180	57 055	7125	11
Norway	7486	6443	1043	14
Australia	34 211	30 440	3771	11
Canada	42 626	39 162	3464	8
Finland	78 175	65 062	13 113	17
England and Wales	65 234	58 962	6272	10
Scotland	6891	6009	882	13
New Zealand	7319	6423	896	12
USA	301 181	253 367	47 814	16
France	138 713	120 494	18 219	13
Switzerland	22 000	19 800	2200	10
Austria	20 884	18 813	2071	10
Spain	22 036	19 015	3021	14
Romania	7105	6759	346	5
Slovakia	3832	3507	325	9

Source: European Federation of National Associations of Orthopeadic and Traumatology (EFORT) (2010).

Throughout the twentieth and into the twenty-first century, a deeper understanding of joint biomechanics has developed. New engineering materials have been produced. Methods for their manufacture into useable products continue to evolve, often aided by computer technology. While these latter trends have been aimed primarily at manufacturing industry, they can also be applied in the engineering of joint replacements. This realization has spurred the major advancements needed to deal with the problems encountered mainly by an ageing population, to the extent that virtually every one of the parts of the human skeleton can now have an industrially produced substitute.

The steps leading to the decision to replace a joint are clear and logical. A main indication is pain, together with impaired movement. An understanding of human anatomy, in particular that of the joints, is needed to evaluate movement, or gait, and select methods of internal inspection. These topics form the substance of the next three chapters.

References

Amin, A.K., Clayton, R.A., Patton, J.T. *et al.* (2006) Total knee replacement in morbidly obese patients. *Journal of Bone and Joint Surgery* **88** (10), 1321–1326.

BBC (2008) *Obesity: In Statistics*, news.bbc.co.uk/1/hi/health/7151813.stm (accessed 23 April 2013).

Busija, L., Hollingsworth, B., Buchbinder, R. and Osborne, R.H. (2007) Role of age, sex, and obesity in the higher prevalence of arthritis among lower socioeconomic groups: a population-based survey. *Arthritis Care and Research* **57** (4), 553–561.

Changulani, M., Kalairajah, Y., Peel, T. and Field, R.E. (2008) The relationship between obesity and the age at which hip and knee replacement is undertaken. *Journal of Bone and Joint Surgery* **90-B** (3), 360–363.

European Federation of National Associations of Orthopeadic and Traumatology (EFORT) (2010) European Arthroplasty Register (EAR). European Arthroplasty Register (EAR) Publications. Innsbruck, Germany.

Horan, F. (2006) Obesity and joint replacement. *Journal of Bone and Joint Surgery* **88-B** (10), 1269–1271.

National Health Service (NHS) (2000) *Information Centre for Health and Social Care*, Office for National Statistics, London.

National Health Service (NHS) National Services Scotland (2012). *Scottish Arthroplasty Project-Biennial Report*, Information Services Division (ISD) Scotland Publications, Edinburgh.

Williams, G. and Fruhbeck, G. (2009) *Obesity: Science to Practice*, Wiley-Blackwell, Chichester, p. 230.

World Health Organisation (WHO) (2003) *Global Strategy on Diet, Physical Activity and Health. Obesity and Overweight Fact Sheet*, World Health Organisation Press, Geneva, Switzerland.

2

Basic Anatomy

2.1 Terminology

The human body consists of three main components: head, trunk and limbs. The trunk is made up of the neck, chest (or thorax) and abdomen (belly). The lower region of the abdomen is the pelvis. The 'perineum' is the lowest part of the pelvis and of the trunk. The central axis of the 'vertebral column', or spine, and the upper (cervical) part of the spine support the head.

The upper and lower regions of the limbs are made up of respectively (i) the arm, forearm and hand, and (ii) thigh, leg and foot.

This structure of the human body and the standard anatomical position are shown in Figure 2.1. The body is standing upright; the feet are together; the head and eyes look to the front. The arms are by the sides of the body. The palms of the hands face forward.

As indicated in Figure 2.1, an imaginary plane is drawn vertically through the middle of the body, from front to back. This is termed the 'median sagittal plane', thus dividing it into right and left halves. The terms 'medial' and 'lateral' are used to describe respectively parts of the body closer to, and further from, the median plane. Alternative expressions are sometimes used. For example, 'ulnar' and 'radial' can replace respectively 'medial' and 'lateral'. In descriptions of the forearm, which has two bones, the radial term can describe the radius on the 'thumb' side or lateral region. The term ulnar can be used to indicate the body part which is medial. For the two bones of the lower leg, the terms 'fibular' and 'tibial' are often used to describe the fibula on the lateral, and tibia on the medial, sides.

The Engineering of Human Joint Replacements, First Edition. J.A. McGeough.
© 2013 John Wiley & Sons, Ltd. Published 2013 by John Wiley & Sons, Ltd.

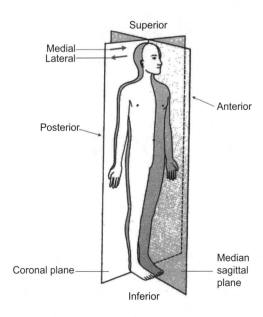

Figure 2.1 Terminology for human body.
Source: Adapted from McMinn, C.M.H., Hutchings, R.T., Pegington, J. and Abrahams, P.H. (1993) *A Colour Atlas of Human Anatomy*, 3rd edn, Wolfe Publishing, London.

Regions nearer to the front of the body are termed 'anterior' (or ventral). Those nearer to the back are denoted 'posterior' (or dorsal). However, for the hand, the anterior surface is described as the 'palm'. Its posterior surface is called the 'dorsum'. The upper surface of the foot is its dorsum, or dorsal surface. The sole of the foot is the plantar surface.

The upper regions of the body are called 'superior' whereas the lower areas are 'inferior'. The terms 'proximal' and 'distal' are used to describe positions respectively nearer to, and further from, the root of the skeletal structure.

A 'sagittal plane' is any plane that is parallel to the median sagittal plane. A 'coronal' (or 'frontal') plane is any plane which is both vertical and perpendicular to the median sagittal plane (McMinn *et al.* 1993).

2.2 Human Skeleton

The structure of the skeleton of the human body is shown in Figure 2.2. The key that accompanies this figure indicates by number the principal parts of the skeleton, and in particular those encountered in joint replacement.

Key:

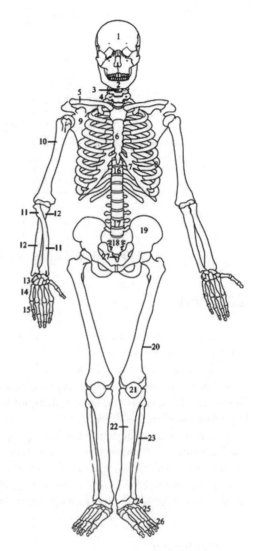

1   Skull

2   Mandible

3   Hyoid bone

4   Cervical vertebrae

5   Clavicle

6   Sternum

7   Costal cartilages

8   Ribs

9   Scapula

10 Humerus

11 Radius

12 Ulna

13 Carpal bones

14 Metacarpal bones

15 Phalanges of thumb and fingers

16 Thoracic vertebrae

17 Lumbar vertebrae

18 Sacrum

19 Hip bone

20 Femur

21 Patella

22 Tibia

23 Fibula

24 Tarsal bones

25 Metatarsal bones

26 Phalanges of toes

27 Coccyx

Figure 2.2   Human female skeleton.
Source: Adapted from McMinn, C.M.H., Hutchings, R.T., Pegington, J. and Abrahams, P.H. (1993) *A Colour Atlas of Human Anatomy,* 3rd edn, Wolfe Publishing, London.

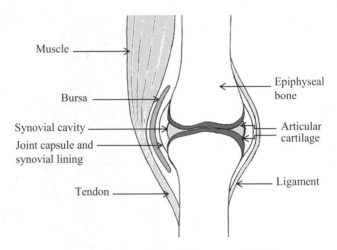

Muscle

Bursa

Synovial cavity

Joint capsule and
synovial lining

Tendon

Epiphyseal
bone

Articular
cartilage

Ligament

Figure 2.3   Structure of the synovial joint.

2.3 Joints

The junction of two bones is termed an articulation or joint. A common purpose of all joints is to hold together the relevant parts of the body. Joints are either fixed or moveable. Moveable or 'diarthrodial' joints enable various motions of the skeleton. Three types of joint are present in the human body: fibrous, cartilaginous and synovial. Synovial joint replacements are the main type discussed in this book.

Synovial joints depend on the properties of articular cartilage, a load-bearing and wear-resistant connective tissue that covers the bones associated with the joint, and on synovial fluid, a nutrient and lubricant within the region of the joint. Both of these substances will be discussed more fully below. Figure 2.3 illustrates the structure of a synovial joint.

2.4 Cartilage

Cartilage contains a type of biological cell called a 'chondroblast'. Chondroblasts produce a cartilaginous matrix composed of collagen, elastin fibres, and a ground substance rich in proteoglycans. Chondroblasts that become caught in spaces in the extracellular matrix termed 'lacunae' are called 'chondrocytes'. Cartilage is avascular (that is, it does not have a blood supply), with the exception of epiphyseal cartilage which is present in the growth plates of long bones.

Key:
O: oxygen;
H: hydrogen
C: carbon;
N: nitrogen;
R: side group;
n: repeating structure.

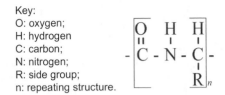

Figure 2.4   Generic arrangement of a polypeptide.

2.5 Protein and Collagen

Proteins are biochemical substances made of polypeptides in a fibrous or globular form. They enable a biological function to proceed. A polypeptide is a single linear polymeric chain of amino acids. The peptide bonds within a polypeptide are formed by the condensation reaction between the amino group of one amino acid and the carboxyl groups of the next amino acid.

Amino acids are differentiated by their side group, denoted R. In the case of glycine – the simplest amino acid – R is a hydrogen atom. Figure 2.4 shows the arrangement of a polypeptide.

Collagen is a structural protein, the main constituents of which are given in Table 2.1. It has the general amino acid sequence of -glycine-proline-hydroxyproline-glycine-X- where X can be any other amino acid. It is arranged in a triple α helix glycine (Gly), shown in Figure 2.5 (a) as a flat sheet structure, the repeating distance of which is 0.72 nm.

For relatively larger side groups, the structure takes the form of a helix. Then the hydrogen bonds bind the helix together, as illustrated in Figure 2.5 (b).

These amino acids are the starting-point in the formation of collagen, as represented in Figure 2.6.

Table 2.1   Constituents of collagen.

Amino acids	Content (mol/100 mol amino acids)
Glycine	31.4–33.8
Proline	11.7–13.8
Hydroxyproline	9.4–10.2
Acid polar (Aspartic, Glutamic, Asparagine)	11.5–12.5
Basic polar (Lysine, Arginine, Histidine)	8.5–8.9
Other	Residue

Source: Adapted from Park, J.B. and Lakes, R.S. (2007) *Biomaterials: An Introduction,* 3rd edn, Springer, New York.

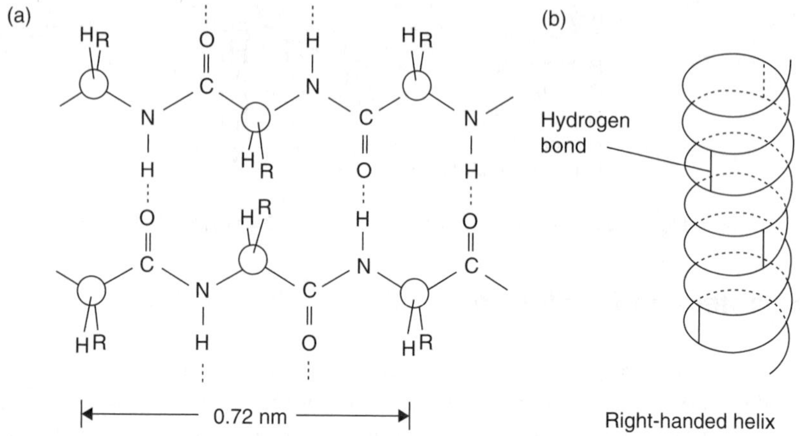

Figure 2.5 (a) Flat sheet structure of protein; (b) helical protein chain (right handed).
Source: Reproduced from Park, J.B. and Lakes, R.S. (2007) *Biomaterials: An Introduction,* 3rd edn, Springer, New York.

They link together producing a molecular chain, which coils into a left-handed helix. The intertwining of three chains in a triple-stranded helix leads to a 'tropocollagen' molecule. On the alignment and partial overlapping of many tropocollagen molecules, collagen fibrils of diameter 20 to 40 nm are produced, bundles of which, with a diameter 0.2–1.2 μm, are the basis of connective tissue.

Tissues containing significant amounts of collagen, such as human cartilage, may have significant load-bearing capacity and tensile strength. Table 2.2 shows the mechanical properties of collagen in relation to elastic fibres. The most common type in the body is the smooth, glossy hyaline cartilage, which contains chondrocytes and type II collagen. It is strong and compressible. The term 'articular' refers to hyaline cartilage, which covers and protects adjacent ends of bones.

Articular cartilage resembles a viscoelastic bearing surface. The structure of the collagen fibril orientation in the cartilage can convert shear forces on the articular surface to compressive forces at the interface between bone and cartilage. The coefficient of friction of cartilage is much less than that of man-made materials, such as Teflon on Teflon. This low coefficient of friction becomes relevant when the kinematics of the knee are considered, as discussed later.

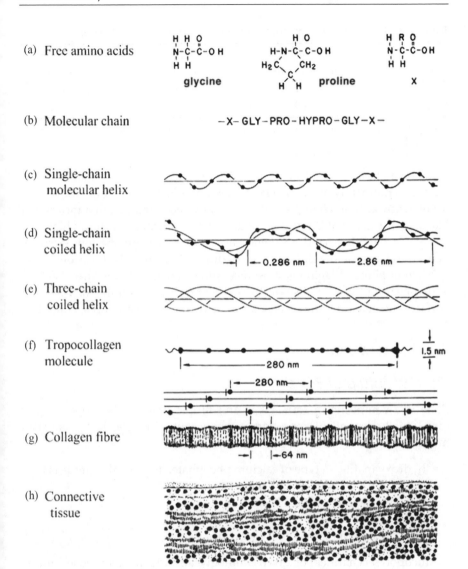

Figure 2.6 Formation of collagen.
Source: Reproduced from Gross, J. (1961) Collagen. *Scientific American* **204**, 121–130.

Table 2.2   Mechanical properties of collagenous and elastin fibres.

Fibre	Modulus of elasticity (MPa)	Tensile strength (MPa)	Ultimate elongation (%)
Elastin	0.6	1	100
Collagenous	1000	50–100	10

Source: After Park, J.B. and Lakes, R.S. (2007) *Biomaterials: An Introduction,* 3rd edn, Springer, New York.

Synovial fluid is a lubricant that bathes the articular cartilage and is described more fully by Balazs (1974). It is a non-Newtonian fluid: the viscosity of synovial fluid changes with shear rates. For example, in a movement like a heel strike (that is impulse loading) the synovial fluid shows high viscosity. On the other hand, the synovial fluid can provide the low viscosity needed to lubricate the sliding surfaces during flexion of the knee joint.

Another fibrocartilaginous substance is the meniscus. In the knee it acts as a load-bearing 'washer'. Its effect is reduction in stress on the load-bearing surfaces by increasing the contact area between femoral and tibial condyles.

2.6 Human Bone

2.6.1 Structure of Bone

Bone is a composite material. It is anisotropic (that is, its properties are directionally dependent), non-homogeneous, and visco-elastic. It comprises mainly:

- hydroxyapatite – a type of calcium phosphate. Its general formula is Ca_{10} $(PO_4)_6$ $(OH)_2$ (about 43% by weight);
- collagen (type 1) (about 36%);
- water (about 14%);
- a small amount of mucopolysaccarides.

Mucopolysaccharides, or glycosaminoglycans, are an unbranched linear chain of repeating subunits. The subunit is hexose, which is a six-carbon sugar or hexuronic acid that is linked to hexosamine (six-carbon sugar that contains nitrogen). Types of glycosaminoglycan can be found in synovial fluid and in connective tissue, cartilage and tendons.

Bone also contains organic material including blood and lymph vessels, nerves, and cells termed osteoblasts and osteoclasts. The latter respectively produce and resorb bone material.

Table 2.3   Constituents of wet cortical bone.

Constituent	Percentage by weight
Mineral (apatite)	69
Organic matrix	22
Collagen	(90–96% of organic matrix)
Others	(4–10% of organic matrix)
Water	9

Source: Adapted from Tiffit, J.T. (1980) The organic matrix of bone tissue, in *Fundamental and Clinical Bone Physiology* (ed. M.R. Urist), J.B. Lippincott, Philadelphia, PA, Ch. 3.

The two main types of bone are cortical (or 'compact') and cancellous (or 'trabecular'). Cortical bone is a hard and compact tissue. The constituents of wet (not dehydrated) cortical bone are given in Table 2.3. (Note that dry bone will exhibit different qualities: for example their densities vary from about 1990 (wet) to 140–1110 (dry) kg/m^3.) Apatite in cortical bone has a similar crystal structure to hydroxyapatite, and is formed in needle-like shapes, 20–40 nm in length, 1.5–3.0 nm in thickness, in the matrix of collagen fibres.

The tissue of cortical bone is composed of secondary osteons, of diameter 100 to 200 μm, each of which is made of concentric lamellae, of thickness about 1 to 5 μm. The lamellae of osteons surround concentrically 'Haversian' canals. The collagen fibres take the shape of a lamellar sheet about 3 to 7 μm thick, which runs in a helical direction with respect to the long axis of the cylindrical system of osteons, or 'Haversians'. The interconnected pores of the Haversian canals, lacunae (cavities which contains osteocytes), or canaliculi (small channels connecting the lacunae), are connected in turn to the bone marrow cavity and enable the transport of metabolic substances. This canal system is filled with body fluid (which may be regarded as essentially water containing solutes such as proteins), of volume about 18 to 19%.

Cancellous bone is more porous than cortical bone. It consists of a continuous three-dimensional network of interconnected rods and plates, known as trabeculae, with cavities that are filled with a viscous fluid (viscosity of approximately 0.04 to 0.4 Pa).

Bone can be classified by its shape. 'Long' bones predominate in the extremities of the human skeleton, and provide levers for movement. In the upper limbs these bones are mainly used for movement such as reaching, grasping and throwing. They are lighter and smaller than those in the lower limbs. The latter bones have to be larger and stronger in order to bear the weight of the body during movement, and associated repeated stress. Figures 2.7 and 2.8

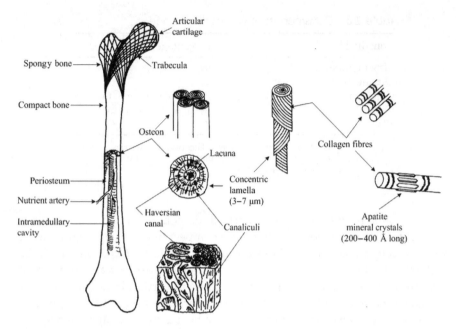

Figure 2.7 Microstructure of long bone.
Source: Reproduced from Park, J.B. and Lakes. R.S. (2007) *Biomaterials: An Introduction,* 3rd edn, Springer, New York.

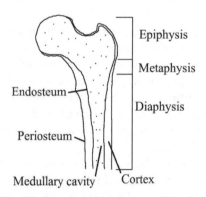

Figure 2.8 Structure of long bone.

show further features of a typical long bone. Its shaft is the 'diaphysis', with an outer shell, the 'cortex', which is made of cortical bone. The outer surface is called the periosteum (and the inner, endosteum). The cortex envelopes bone marrow contained within the medullary cavity. An 'epiphyseal' plate of cartilage separates the metaphysis (wider portion of long bone which grows and lengthens during childhood) from the epiphysis. The epiphysis also contains trabecular bone, and marks the proximal and distal bounds of the long bone.

'Shorter' bones are present in the hands and feet. Like their longer counterparts they also provide for movement. In addition they supply elasticity, flexibility and shock absorption. Other 'flat' bones, for example in the pelvis and scapula, protect underlying structures and provide locations for attachment of muscles.

The vertebral column is made up of irregular bones, part of whose function is to provide for muscular attachment. The vertebral bone also absorbs the impact forces associated with walking, running and jumping.

From Figure 2.9, key features of bone needed in understanding joint replacement anatomy can now be identified. A 'condyle' is a rounded part of bone that articulates with another bone, an 'epicondyle' being a smaller version. A facet is a small, flat, smooth bone where articulation occurs. The

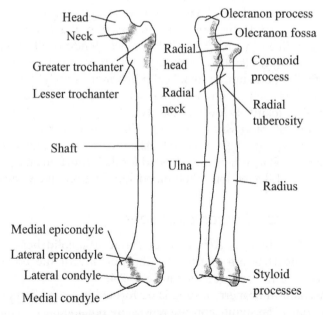

Figure 2.9 Main features of long bone.

fossa part of a bone is shaped like a shallow dish. It gives space for the articulation of bone or for attachment of muscles. A 'process' is a prominence in bone that is often complementary to a depression in another bone. Muscles and tendons can attach to raised sections of bones called 'tuberosities' or 'apophyses'.

Both muscle and tendon tissue play a major part in enabling the range of motion of a joint. Muscle use enables new bone cells to be set down to create or raise the apophysis.

Muscles are made of bundles of cellular fibres, 'myofibrils'. Human muscle is of two kinds: voluntary (striated) regulated by the central nervous system, and involuntary, regulated by the autonomic nervous system, not by the brain. Joint surgery demands a careful consideration of the function of the muscles relating to an articulation so that as much normal movement is possible after a joint defect has been corrected.

Tendons are made up of connective tissue, which attaches the muscles to the bone. They are often the focus of surgery, sometimes after they have been torn or cut, sometimes as a part of an operation for articular reconstruction when the insertion of tendons, often within a bone, is carefully considered. The structure of tendons centres on the number and properties of the cells that manufacture the intercellular matrix. The principal macromolecules that form the matrix are collagen (mainly type I) and proteoglycan. In the case of the knee, for example, the principal component of the quadriceps tendon is collagen, arranged in strong parallel bundles. Tendons, like menisci and discs, have low water content. Although rich in collagen, tendons have a relatively small blood supply; they heal much more slowly than bones like the tibia, which is composed of blood vessel-rich cancellous bone as well as a dense cortex of relatively low vascularity.

Ligaments are fibrous tissue containing collagen which connect one bone to another to form a joint. They transmit tensile forces. They can vary greatly in size and shape. Collagen rich, they depend for their function on a population of proteoglycans. Like tendons, their blood supply is relatively small.

2.6.2 Mechanical Properties of Bone

Although cortical bone is composed of about 90% solid bone tissue, its mechanical properties are affected by the porosity of the bone. Typical mechanical properties are presented in Table 2.4.

When mechanical property testing is performed in a direction perpendicular to the bone axis, significant differences are obtained, as exemplified in Table 2.5. The magnitude of the elastic modulus of bone is affected by many

Table 2.4 Typical mechanical properties of human bones (direction of test is longitudinal to bone axis).

Bone	Modulus of elasticity (GPa)	Tensile strength (MPa)	Compressive strength (MPa)
Leg			
Femur	17.2	121	167
Tibia	18.1	140	159
Fibula	18.6	146	123
Arm			
Humerus	17.2	130	132
Radius	18.6	149	114
Ulna	18.0	148	117

Source: From Yamada H. (1970) *Strength of Biological Materials,* Williams & Wilkins Publishers, Baltimore, MD.

conditions, including its porosity, density, mineral content, and lamellar orientation. McCalden *et al.* (1993) show how the tensile properties of cortical bone decrease with age, drawing attention to the greater decline in strength with female bone.

The hydraulic effects arising from this interaction of fluid and solid facilitates the load-bearing capacity of cancellous bone. The mechanical behaviour of cancellous bone may therefore be considered to be viscoelastic, being effectively elastic for normal strain rates that bone meets (about 1 Hz).

However, the elastic modulus of cancellous bone is anisotropic, and the magnitude of its other mechanical properties depends on the bone structure. Typical mechanical properties for cancellous bone are presented in Table 2.6.

McCalden *et al.* (1997) show how the compressive strength of human cancellous bone decreases with age. Little data are available on its other visco-elastic mechanical properties such as creep and fatigue which are time

Table 2.5 Mechanical properties of human long bone: testing parallel and perpendicular to bone axis.

Property	Parallel to bone axis	Perpendicular to bone axis
Elastic modulus, (GPa)	17.4	11.7
Ultimate tensile strength (MPa)	135	61.8
Poisson's ratio	0.46	0.58
Elongation at fracture (%)	3–4	–

Source: After Gibbons, D.F. (1976) Biomedical materials, in *Handbook of Engineering in Medicine and Biology* (eds D.G. Fleming and B.N. Feinberg), CRC Press, Boca Raton, FL, pp. 253–254.

Table 2.6   Mechanical properties of cancellous bone.

Apparent density (kg/m^3)	Elastic modulus (MPa)	Poisson's ratio
140–1110 (Average 620)	10–1570	0.4

Source: After Keaveny, T.M. (1998) Cancellous bone, in *Handbook of Biomaterial Properties* (eds J. Black and G. Hastings), Chapman & Hall Ltd, London, U.K.

dependent. (In most analyses its mechanical properties are considered to be independent of time.) The hydraulic permeability of cancellous bone becomes a significant property when bone cement is used to fix artificial joint replacements.

2.6.3 Bases of Biomechanics of Joints

These properties and characteristics of bone and its supporting tissue enable the biomechanical behaviour of synovial joints to be appreciated. Friction between moving bones is decreased by the articular cartilage, which can also absorb any effects of shock or abrupt movement. A 'capsule' containing fibrous tissue and surrounding the joint, together with ligaments, holds it in place. This capsule is lined with a membrane which secretes synovial fluid. The presence of synovial fluid in the cavity between the articular cartilages reduces the effect of friction. Ready movement of the joint should then be achieved (Figure 2.10).

These movements for the main joints that may require replacement can now be summarized. The hip and shoulder joints take the form of a ball and socket (see Figure 2.10 (d)). The head of the femur is rounded. It fits into the cup-shaped socket of the pelvic bone. This configuration enables the hip joint to move in most directions.

The elbow, knee, and interphalangeal (finger) joints resemble a hinge. Movement occurs in a single plane in a backward and forward direction. As indicated in Figure 2.10 (a), the cylindrical end of one bone nests within the curved recess of the other bone of the joint.

The wrist joint is condyloid in shape: one end of its main bone has an egglike shape, with the other featuring an oval recess (see Figure 2.10 (b)). This configuration enables the wrist to bend and straighten and to move in a side-to-side action. The wrist also possesses sliding joints (in its carpals, see Chapter 3). These arise from flat articular surfaces between its bones. They enable a sliding action to be performed. This movement is restricted by the presence of strong ligaments.

The joint at the base of the thumb is a saddle joint, which features two articular U-shaped surfaces (see Figure 2.10 (c)). The joint moves in two

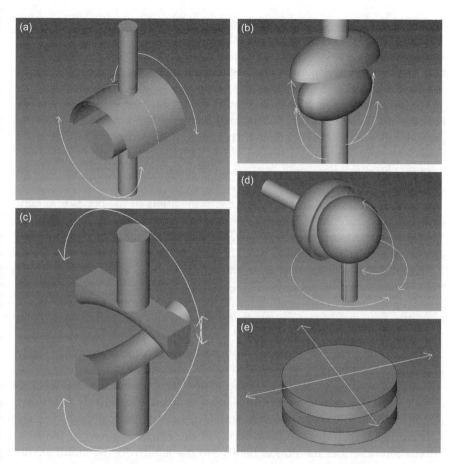

Figure 2.10 Main movements of joints: (a) hinge, (b) condyloid, (c) saddle, (d) ball and socket, and (e) gliding.

planes, for example it can move across the palm of the hand, and also touch the tips of the fingers.

The heel joints at the tarsals possess gliding motion, similar to two plates sliding over each other, as shown in Figure 2.10 (e). Other joints in the foot function differently. For example, the ankle and toe joints present themselves as hinges (Abrahams *et al.* 2005).

In humans, movement is of two kinds: linear and spiral (helical) (Pettigrew 1873). The role of human bones in accommodating linear movement is analogous with the movement of a crane, a car, or a bicycle. Helical movements are most readily seen in swimming, exemplified by the 'crawl' and the 'backstroke'. They can be represented by the structure of, for example, the elbow, which has to facilitate both linear and helical movement. A structure to allow helical movement is also seen in the lower leg where 'twisting' is part of daily walking.

The term 'flexion' describes the position made possible by decrease in angle between the bones of a joint. For instance when the hand moves closer to the shoulder the elbow becomes flexed. This flexed position of the elbow is made possible by the combined action of bones, cartilage, ligaments, muscles and tendons. Flexion arises only in the sagittal plane, that is, in a forward to backward direction.

The term 'extension' implies increase in the angle in the bones of a joint causing it to straighten. These expressions and variations of them that are special to some joints are used in describing movements of the shoulder, elbow, wrist, phalangeal joints, hip, knee and ankle.

Movement of a distal part from the midline of the body in the coronal plane is 'abduction'. That towards the midline is 'adduction'. 'Rotation' can be either 'internal or external'. Internal rotation makes the anterior surface of a distal segment rotate medially relative to the proximal segment. In 'external rotation' the anterior surface of a distal segment rotates laterally relative to the proximal segment. The positions of joints can be internally and externally rotated.

With the wrist joint the terms 'supination' and 'pronation' are used to describe the positions for which the palm is facing respectively anteriorly and posteriorly, with the arm in the standard anatomical position. The same two terms are included in descriptions of the planar motions of the foot. Supination describes its motion relative to the leg involving simultaneous flexion, internal rotation and inversion. Inversion is motion in the coronal plane with the plantar surface facing in the medial direction. Pronation is used to describe the corresponding foot motion with simultaneous flexion, external rotation and 'eversion'. Eversion is the motion of the foot in the coronal plane with the plantar surface facing in the lateral direction.

These main joints and their movements are further described in the next chapter. Further useful information on the properties of bone can be found in Currey (2006) and Bonucci (2000).

References

Abrahams, P., Craven, J. and Lumley, J. (2005) *Illustrated Clinical Anatomy*, Hodder Arnold, London.

Balazs, E.A. (1974) The physical properties of synovial fluid and the special role of hyaluronic acid, in *Disorders of the Knee* (ed A. Helfet), T.B. Lippincott Company, Philadelphia, pp. 63–75.

Bonucci, E. (2000) Basic composition and structure of bone, in *Mechanical Testing of Bone and Bone-Implant Interface* (eds Y.H. An and R.A. Draughtan), CRC Press, Washington DC.

Currey, J.D. (ed.) (2006) *Bones: Structure and Mechanics*, Princeton University Press, Princeton, NJ.

Fleming, D.G. and Feinberg, B.N. (1976) *Handbook of Engineering in Medicine and Biology*, CRC Press, Boca Raton, FL.

Gibbons, D.F. (1976) Biomedical materials, in *Handbook of Engineering in Medicine and Biology* (eds D.G. Fleming and B.N. Feinberg), CRC Press, Boca Raton, FL, pp. 253–254.

Gross, J. (1961) Collagen. *Scientific American* **204**, 121–130.

Keaveny, T.M. (1998) Cancellous bone, in *Handbook of Biomaterial Properties* (eds J. Black and G. Hastings), Chapman & Hall Ltd, London, U.K.

McCalden, R.W., McGeough, J.A., Barker, M.B. and Court-Brown, C.M. (1993) Age-related changes in the tensile properties of cortical bone. *Journal of Bone and Joint Surgery* **75-A** (8), 1193–1205.

McCalden, R., McGeough, J.A. and Court-Brown, C. (1997) Age-related changes in compressive strength of cancellous bone. The relative importance of changes in density and trabecular architecture. *Journal of Bone and Joint Surgery* **79-A** (3), 421–425.

McMinn, C.M.H., Hutchings, R.T., Pegington, J. and Abrahams, P.H. (1993) *A Colour Atlas of Human Anatomy*, 3rd edn, Wolfe Publishing, London.

Park, J.B. and Lakes, R.S. (2007) *Biomaterials: An Introduction*, 3rd edn, Springer, New York.

Pettigrew, J.B. (1873) *Animal Locomotion or Walking, Swimming, and Flying*, Henry S. King & Co., London.

Tiffit, J.T. (1980) The organic matrix of bone tissue, in *Fundamental and Clinical Bone Physiology* (ed. M.R. Urist), J.B. Lippincott, Philadelphia, PA, Ch. 3.

Yamada, H. (1970) *Strength of Biological Materials*, Williams & Wilkins Publishers, Baltimore, MD.

3

Anatomy of Joints

3.1 Shoulder

3.1.1 Anatomy of the Shoulder Joint

The shoulder comprises three bones: the clavicle (collar bone), scapula (shoulder blade) and humerus (upper arm bone), all of which have associated muscles, ligaments, tendons, blood vessels, nerves, and lymphatics.

Figure 3.1 and 3.2 show the main features of the shoulder joint.

The 'glenohumeral' joint is the main shoulder joint. It takes the form of a ball and socket. The former component is formed from the humeral rounded medial anterior surface; the latter is made up of a shallow dishlike part of the lateral scapula, or 'glenoid fossa'. Owing to the shallow aspect of the fossa and also as the shoulder is comparatively flexible in its connection to the body, the arm is able to rotate circularly, and to act like a hinge moving up and away from the body. The glenohumeral joint has an envelope of soft tissue termed the 'capsule' which is attached to the scapula, humerus and the head of the biceps. Like all synovial joints, it has a thin, smooth synovial membrane. The strength of the articulation is assisted by the 'coracohumeral' ligament, which attaches parts of the scapula to the greater tubercle humerus. Another three 'glenohumeral' ligaments attach the lesser tubercle of the humerus to the lateral scapula.

The shoulder also features the 'sternoclavicular' joint. It is situated at the medial end of the clavicle, a triangularly shaped and rounded bone, with the convex-shaped 'manubrium' bone, at the superior end of the sternum.

The Engineering of Human Joint Replacements, First Edition. J.A. McGeough.
© 2013 John Wiley & Sons, Ltd. Published 2013 by John Wiley & Sons, Ltd.

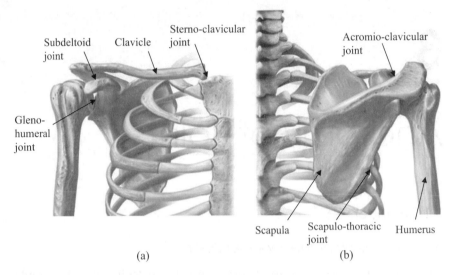

(a)   (b)

Figure 3.1 Shoulder joints (a) anterior view (b) posterior view.
Source: Reproduced from Tortora, G.J. and Derrickson, B.H. (2012) *Principles of Anatomy and Physiology*, John Wiley & Sons, Inc., New York.

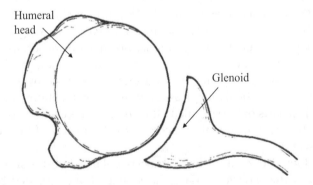

Figure 3.2 Shoulder joint: glenoid and humeral head.
Source: Adapted from Wallace, W.A. (ed.) (1998) *Joint Replacement in the Shoulder and Elbow*, Butterworth Heinemann, Oxford.

These two bones can articulate. Stability of the joint is obtained through its tight capsule and an intra-articular disc, movement being limited by the costoclavicular ligament. Nevertheless the presence of a fibrocartilaginous disc at the joint facilitates increase in the range of movement.

In common with all synovial joints, the ability of the bones of the shoulder joint to articulate is facilitated by the articular cartilage on the ends of the ball and socket of the joint. A second type of cartilage, 'labrum', which is more fibrous and rigid, is attached around the socket of the joint. The 'acromioclavicular' joint is the third such component of the shoulder.

The locations of the scapula and clavicle are noted in Figures 3.1(a) and 3.1(b). They constitute the 'pectoral girdle'. The pectoral girdle connects the upper limbs to the axial skeleton; it provides stable support for the glenoid. The latter articulates with the humeral head, as illustrated in Figure 3.2. The muscles that facilitate movement of the shoulder are attached to the scapula, clavicle and humerus. There are two main types, 'movers' and 'humeral head stabilizers', which are referred to below.

On movement of the upper limb, the pectoral girdle itself moves, with motion taking place in the following locations:

- movement of the scapula causes some motion of the sternoclavicular joint;
- acromioclavicular joint, which is stabilized mainly by the coracoclavicular ligaments, allowing movement between the clavicle and the scapula;
- 'muscle sliding' scapulothoracic joint allows the scapula, which is attached to the chest, to slide over the underlying rib cage;
- the glenohumeral joint is responsible for two-thirds of the total movement at the shoulder, and is the main articulation;
- subacromial joint, which permits a free sliding movement.

These movements enable the hands and arms to undertake their normal actions, and permit more strenuous activity including pushing, pulling and lifting.

3.1.2 Biomechanics of the Shoulder Joint

These movements may be more fully interpreted from an examination of the kinematics of the shoulder.

The upper limb including the shoulder accounts for about 5.2% of the entire body weight – that is, about 3.6 kg for a man weighing 70 kg. For a

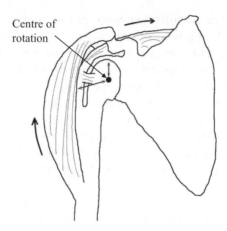

Figure 3.3   Indication of lever arms from centre of rotation at shoulder on elevation of arm (arrows represent direction of movement).
Source: Adapted from Wallace, W.A. (ed.) (1998) *Joint Replacement in the Shoulder and Elbow,* Butterworth Heinemann, Oxford.

person in the standing position, this weight counteracts the upward force of the 'deltoid' muscle. The latter is one of the 'mover' muscles placed around the shoulder joint. Their general effect is to abduct, adduct, flex, rotate and extend, providing elevation, internal and external rotation, and horizontal flexion and extension.

The movement of elevation is the plane in which the arm can be raised to its highest position (167 and 171° respectively for men and women, although arthritis can limit this angle). For the standing position, it is the angle between the upper arm (or humerus) and the vertical line that passes through the shoulder joint.

The level arms involved in the shoulder joint on elevation of the arm are indicated in Figure 3.3. Wallace (1998) comments that elevation is a more functional action compared, for example, to abduction in the plane of the scapula.

Wallace calculates that for a 70 kg male whose arm is held at a 90° angle, and a 28 mm lever arm, the deltoid needs a force of 43 kg to maintain this position. On the addition of a 2 kg weight to the hand, the force required by the deltoid rises to 87 kg.

Elevation of the arm will lead to other movements such as the external rotation of the humerus, as indicated in Figure 3.4.

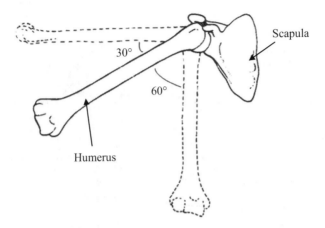

Figure 3.4   External rotation of the humerus on elevation of the arm.
Source: Adapted from Wallace, W.A. (ed.) (1998) *Joint Replacement in the Shoulder and Elbow,* Butterworth Heinemann, Oxford.

3.2 Elbow

The elbow has to flex and extend through large ranges of motion to control the distance of the hand from the body. As it must also enable forearm helical movements and rotation, associated with other motions required for the hand, it possesses three types of joint.

3.2.1 Anatomy of the Elbow Joint

The joints are formed of three bones, namely the humerus of the upper arm and the radius and ulna of the forearm. The 'humeroulnar' joint, which connects the ulna and humerus, is a simple hinge-like joint which enables flexion and extension. The 'humeroradial' joint, which goes from the head of the radius to the humerus, is also a hinge joint. The proximal radioulnar joint is attached between the head of the radius and the radial notch of the ulna. For flexion or extension, the radius carrying the hand can be rotated in this joint, including actions of pronation and supination, that is, rotational movement of the forearm so that the palm faces respectively downwards and upwards.

The bony prominence at the upper end of the ulna is termed the 'olecranon'; the inner aspect of the elbow is the 'antecubital fossa'.

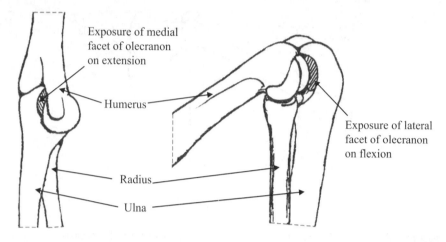

Figure 3.5 Main features of the elbow joint.
Source: Adapted from Wallace, W.A. (ed.) (1998) *Joint Replacement in the Shoulder and Elbow,* Butterworth Heinemann, Oxford.

The articular surfaces of the humerus, ulna and radius are connected by the synovial joint capsule, which is thickened in the medial and lateral regions, and also but less so anteriorly and posteriorly, and takes the form of major ulnar collateral, radial collateral, and annular, ligaments.

Figure 3.5 illustrates the main features of the elbow.

3.2.2 Biomechanics of the Elbow Joint

Flexion and extension of the elbow between the humerus and ulna arise when the elbow undergoes bending and straightening. Figure 3.6 indicates the positions of the humerus for the flexion and extension of the elbow. The line denoted by 'A' shows the longitudinal axis of the humerus; that designated as 'B' indicates the line of attachment of the anterior portion of the medial ligament. In these movements, the elbow is considered to act like a hinge, the movements of which can be interpreted from biomechanics.

When the forearm is turned over, that is pronation or supination, the articulation occurs between the radius and the ulna. (It may be noted that this action can also arise at the wrist.)

As the elbow extends, the forearm flexes in an arc about a single fixed axis that passes through the centres of the capitellum (rounded protuberance at lower end of humerus) and the trochlea (medial portion of the articular

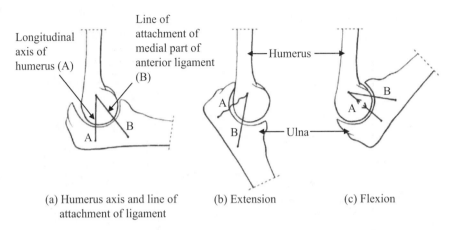

(a) Humerus axis and line of   (b) Extension   (c) Flexion
    attachment of ligament

Figure 3.6   Positions of humerus for common movements of the elbow joint. *Source:* Adapted from Wallace, W.A. (ed.) (1998) *Joint Replacement in the Shoulder and Elbow,* Butterworth Heinemann, Oxford.

surface of the elbow joint). For a fully extended elbow, flexion is defined as $0°$. On flexion, the forearm deviates laterally from the arm, to about respectively $11°$ and $14°$, for respectively male and female adults. Female flexion strength is about 55% of that of the male. The mean range of flexion is $142°$, for male adults. As the elbow extends, its main effectors are the triceps. As it flexes, the main flexors of the elbow are the biceps, brachialis, and brachioradialis (muscles in respectively the upper arm and forearm).

Pulling and pushing are the main actions of the elbow: for the elbow in full extension the associated forces are respectively about 530 N and 590 N (Wallace 1998).

The actions of its adjacent muscles greatly influence the elbow joint forces.

The moment arms for elbow muscles associated with flexion are shown in Figure 3.7, which include the principal flexors of the elbow – the biceps, brachialis and brachioradialis. (note that moment arm is the perpendicular distance from the line of muscle force application to axis of rotation (University of California 2000).

Joint forces active at the elbow during flexion and extension are given in Figure 3.8.

From these calculated results, the forces acting on both the humero-radial and humero-ulnar joints are noted to be greater than those on the hand, when it is close to full extension. This condition arises from the proximity of the

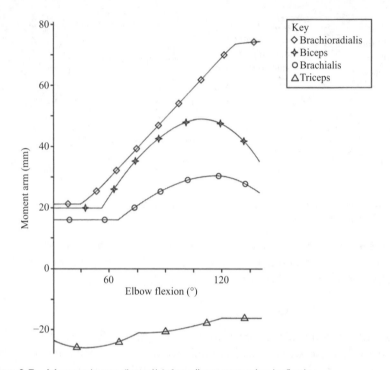

Figure 3.7   Moment arm (length) for elbow muscles in flexion.
Source: After Wallace, W.A. (ed.) (1998) *Joint Replacement in the Shoulder and Elbow*, Butterworth Heinemann, Oxford.

flexor muscles to the axis of flexion of the elbow and is consequently a major mechanical disadvantage that the muscles have to overcome.

Wallace (1998) has explained that when the elbow is extended, the sagittal forces act at the end of the humerus. On flexing the elbow, these forces 'swing round' to act on the anterior aspect. The studies of Souter (1977), from which this analysis of force is drawn, indicate that the elbow behaves as a bicondylar (with meniscus lying between the articular surfaces) joint. The radius and ulna transmit similar loads to the humerus. (When loosening of replacements occurs, the effect of the anterior-posterior component of these forces plays a key part.)

The humeroulnar force is caused by a tension of almost 3 kN arising in the triceps tendon. A large tensile stress then occurs in the bone of the olecranon. If a person falls, the force of stretching the biceps raises the tension, and the olecranon may fracture; even if the bone does not fracture, the impact of the

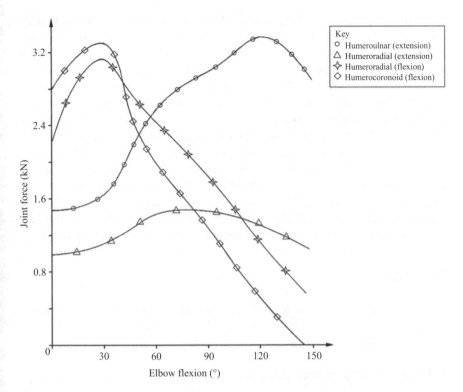

Figure 3.8 Joint forces for elbow joint during flexion and extension.
Source: After Wallace, W.A. (ed.) (1998) *Joint Replacement in the Shoulder and Elbow*, Butterworth Heinemann, Oxford.

posterior aspect of the elbow on the ground may also cause fracture. When the elbow joint is replaced, unnecessary removal of the bone of the olecranon has to be avoided, as RA (rheumatoid arthritis) can also cause erosion.

Other biomechanical actions of the elbow also need consideration. For example, on lifting, large tensile loads occur on the upper limit. The joints of the upper limb are then compressed by the actions that the muscles need to overcome the external load. The resultant compressive force on the radial head and coronoid of the elbow joint line has been estimated to be about 210 N (Wallace 1998). These studies indicate that elbow joint replacements should not be subjected to tensile forces that act in the regions where they have been fixed in place.

The elbow can undergo considerable strain by actions such as pressure of the hands inwards to hold a box in front of the body. This action is known as 'forearm adduction' and the elbow can be flexed to 90°. Poppen and Walker (1978) show that large rotational forces are then likely to occur, with a force of 2.6 N per N of external force estimated to pass through the elbow. Such forces of the magnitude can cause loosening of an elbow replacement.

Another action affecting the elbow arises when the fingers adopt a hook-type grip, for example for tearing apart an object: this is called 'forearm abduction', and the elbow flexes to 90°. Amis (1998) discusses the finger flexor tensions that flex the wrist, as well as the action of the external load. All the major wrist extensors cross the elbow compressing the radial head. He quotes the calculations of forces made by Hunsicker (1955). The mean abduction force is 156 N, the mean adduction force is 218 N, and the humeral external rotation torque is 48 Nm. He calculates a humeroradial force of 0.7 kN, a humero-ulnar force of 1.65 kN, with a wrist force of 2.76 kN. The anatomy and biomechanics of the wrist joint therefore follow in the next section.

3.3 Wrist

The wrist enables the hand to undertake a wide range of everyday actions, providing it with wide mobility and also the strength needed for additional movements such as those associated with lifting and gripping.

3.3.1 Anatomy of the Wrist Joint

Eight small 'carpal' bones are present in the wrist. They form the proximal skeletal region of the hand. The carpals are the connection between the two bones of the forearm, the radius and ulna, with the bones of the hand. The long bones underneath the palm of the hand are the five 'metacarpals', which are attached to the bones of the fingers and thumb, the 'phalanges'.

Figure 3.9 (a) shows these skeletal attributes of the wrist. Part (b) is discussed in section 3.3.2.

The adjoining bones of the right hand are given in Figure 3.10. Each small bone in the wrist forms a joint with its neighbour, being connected through ligaments. The wrist bones are also attached by ligaments to the radius, ulna and metacarpal bones.

The surfaces of the bones are protected by articular cartilage from friction when they rub together and from abrupt impact.

The radiocarpal wrist joint is ellipsoidally shaped. It is formed proximally by the radius and the articular disc, and distally by the proximal row of carpal bones. The joint capsule makes continuous contact with the midcarpal joint, and is strengthened by the presence of ligaments.

(a)

(b)

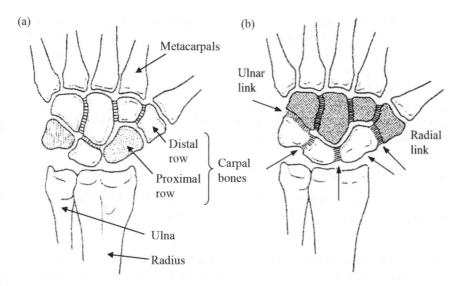

Metacarpals

Ulnar
link

Distal
row

Proximal
row

} Carpal
  bones

Radial
link

Ulna

Radius

Figure 3.9   Skeletal features of the wrist.
Source: Reproduced from Lichtman, D.M., Schneider, J.R., Swafford, A.R. and
Mack, G.R. (1981) Ulnar mid carpal instability-clinical and laboratory analysis.
Journal of Hand Surgery **6** (5), 515–523.

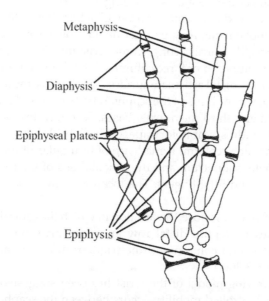

Metaphysis

Diaphysis

Epiphyseal plates

Epiphysis

Figure 3.10   Location of joints of the right hand.

This joint is separated from the distal radioulnar pivotal joint by an articular disc, situated between the radius and the styloid process of the ulna. The latter joint sits between the bones of the forearms, the radius and ulna. The midcarpal S-shaped joint separates the proximal and distal rows of the carpal bones. The intercarpal joints are located between the bones of each row, and are strengthened from the associated ligaments. The intermetacarpal joints between the bases of the metacarpal bones are also strengthened by neighbouring ligaments.

3.3.2 Biomechanics of the Wrist Joint

Youm and Yoon (1979) report that the centres of rotation of the wrist joint are found in the capitate. The joint provides the following main movements for the hand:

- Abduction, movement towards the thumb, and adduction, movement towards the little finger. These movements occur at the radio- and mid-carpal joints.
- Flexion (palmar action of tilting towards the palm), extension (dorsi-flexion, tilting towards the back of the hand).

The flexion actions arise back-to-front about the dorsopalmar axis passing through the capitate bone. As the finger flexors are stronger than the extensors, palmar flexion is a more powerful action than extension.

Other wrist movements may be intermediate between or combine these two main actions. The forces acting on the wrist may be interpreted by considering it to be composed of three columns. Firstly the central force-bearing column includes the distal articulate surface of the radius, the lunate and capitates, the proximal two-thirds of the scaphoid trapezoid, and the articulations with the second and third metacarpal bases. Secondly, the radial column is composed of the radius, scaphoid, trapezoid and the thumb carpometacarpal joint. Finally the ulnar column is made up of triangular fibrocartilage (articular disc), hamate, triquetrum and the articulations of the carpometacarpal joints of the ring and little fingers. The location of these bones is shown in Figure 3.11.

Taleisnik (1985) has proposed a variation in which the central force-bearing column includes the entire distal row and the lunate with the scaphoid included as the lateral column and the triquetrum as a rotary medial column. (See Figure 3.9(a).)

An alternative ring model of the wrist has been suggested by Lichtman (1988). Normal controlled mobility occurs between the scaphotrapezial and triquetrohamate joints. The centre of motion is located in the proximal

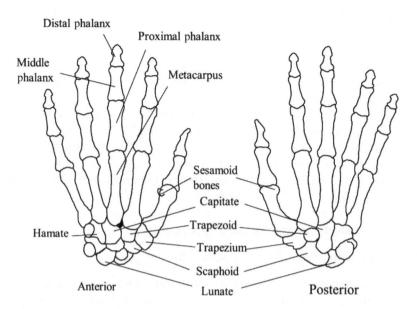

Distal phalanx

Proximal phalanx

Middle phalanx

Metacarpus

Sesamoid bones

Capitate

Hamate

Trapezoid

Trapezium

Scaphoid

Anterior

Lunate

Posterior

Figure 3.11 Bones of right hand.
Source: Adapted from Bowker, P., Condie, D.N., Bader, D.L. and Pratt, D.J. (1993) *Biomechanical Basis of Orthotic Management,* Butterworth Heinemann, Oxford.

capitate. During flexion and extension most motion occurs at the radial carpal joint. (See Figure 3.9 (b).)

The complexity of the kinematics has prompted Ferris *et al.* (2000) to debate whether the wrist functions as a column or as a row of carpal bones. Based on other studies of the range of movement obtained with the wrist, such as those of McQueen *et al.* (1996) and McQueen (1998), they conclude there are two types of wrist movements sometimes with more rotation of the lunate.

Table 3.1 shows typical angular ranges of motion for the wrist.

Table 3.1 Angular ranges (in degrees) of the wrist.

Age (years)	0–19	20–29	30–39	40–49	50–80+
Flexion	85	76	85	82	81
Extension	55	49	51	66	44
Abduction	37	31	33	27	29
Adduction	25	21	24	22	23

Source: Adapted from Bowker, P., Condie, D.N., Bader, D.L. and Pratt, D.J. (1993) *Biomechanical Basis of Orthotic Management*, Butterworth Heinemann, Oxford.

3.4 Finger

3.4.1 Anatomy of the Finger Joints

The anatomical features of fingers are closely related to those of the wrist as discussed in section 3.3 above.

'Extrinsic' refers to the parts of the hand that are connected, for example by ligaments or muscles, to a location outside the hand. The extrinsic hand comprises the proximal carpals which are connected to the radius and ulna. 'Intrinsic' deals with the fingers, which pertain exclusively to the hand. The intrinsic hand consists of the 13 bones that form part of the wrist: the eight distal carpal bones and five metacarpal bones.

3.4.2 Biomechanics of the Finger Joints

The hand is associated with movement at three joints:

- At the carpometacarpal joint of the thumb, movement occurs in five directions: flexion, extension, adduction, abduction, and rotation (of the metacarpal bringing the thumbnail into a plane that is parallel with the palm).
- The metacarpal-phalangeal joints of the thumb and fingers enable flexion-extension movement from 30° (in the thumb) to 90°, and a lesser degree of abduction from, and adduction to, the midline of the middle finger.
- The interphalangeal hingelike movements of the joints of thumb and fingers are restricted to flexion and extension.

For the fingers, the range of flexion is respectively 90° and 45° at the proximal and distal interphalangeal joints. In contrast, at the interphalangeal joint of the thumb, the range of movement is about 80° (Hamblen and Simpson 2010). Further information in the biomechanics of these joints may be found in Unsworth (1981).

3.5 Hip

The hip is one of the four main structural parts of the lower extremity of the body. Its main purposes are to provide locomotion and weight-bearing capacity. Its movement is controlled by at least 17 muscles; acting as flexors, extensors, adductors, and abductors, they enable the main rotational movements of the hip. Posture is controlled by a relationship between the muscles of the lower back and the pelvis.

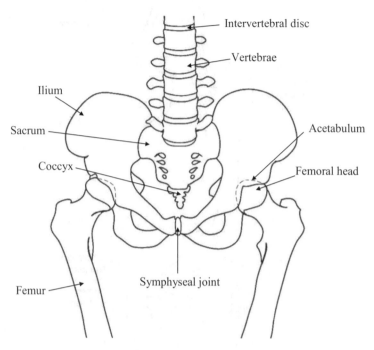

Figure 3.12   Skeletal structure of the hip joint.

3.5.1 Anatomy of the Hip Joint

The hip joint is depicted in Figure 3.12. Its components resemble a ball and socket, sited between the head of the femur (see below) and acetabulum.

The femoral head and the acetabulum are covered by articular cartilage, which should allow painless and smooth motion of the joint. The acetabular articular surface has a horseshoe shape.

The femur and acetabulum are attached by a strong ligament at the femoral head to a notch in the acetabulum. Three thick ligaments encompass the capsules, and limit undue extension of the joint.

The stability of the hip joint is therefore achieved by the acetabulum and its attachment to the femur, the strong capsule and the ligaments that spiral around it, and adjacent muscles.

Figure 3.13 shows further features of the femur, with some of its associated muscle and ligament attachments. The ligament of the head of the femur is accommodated in a pit (the 'fovea') sited in the centre of the head. The neck of the femur is situated primarily at the capsule of the hip joint, and sits at an angle of approximately 125° to its shaft.

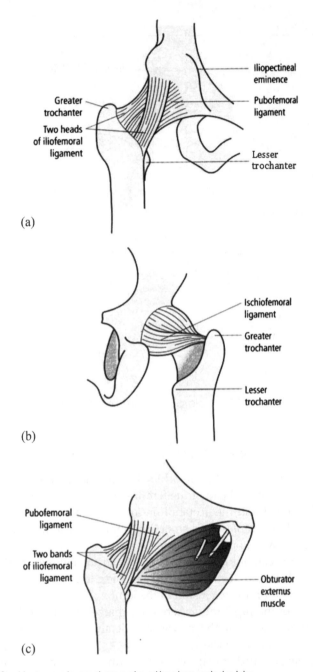

Figure 3.13 Ligaments and muscle attachments to hip.
Source: Reproduced from Abrahams, P.H., Craven, J.L. and Lumley, J.S.P. (2011) *Illustrated Clinical Anatomy,* 2nd edn, Taylor & Francis Ltd., London.

This figure also includes other aspects of anatomy needed in understanding the relationship between the femur and the hip. From Figure 3.13 (a), near to the femoral head is the 'greater trochanter', which extends from the lateral part of femur. The greater trochanter provides attachment to the short rotators of the hip joint and the obturator muscles, indicated in Figure 3.13 (b).

The 'lesser trochanter', a bony protrusion, is situated at the junction of the neck of the femur and its body. It provides attachment to the adductors, biceps and quadriceps, in these regions of the femur. The shaft of the femur slopes in the medial direction at an angle of 10°. These positions of the greater and lesser trochanters with associated ligaments are also shown in Figure 3.13.

At the distal end, the medial and lateral condyles provide articulation with the tibial condyles and the patella of the knee (see section 3.6). At the posterior part, the intercondylar notch separates the articular surfaces of the condyles. In the anterior region these surfaces are united by the concave articulation of the patella. The tibial articular surface of each condyle is convex. Attachment to the medial and lateral ligaments of the knee is secured by the epicondyle, a small elevation on the side of each condyle.

3.5.2 Biomechanics of the Hip Joint

The hip undergoes cyclic loading that is three to five times that of the weight of the body, and has to withstand loads as high as 12 times the body weight. These effects of body weight can be interpreted from Figure 3.14, which illustrates the main forces on the hip.

The body weight is regarded as a load applied to a lever arm extending from the centre of gravity (X) of the body to the centre (B) of the head of the femur. The moment produced by the lever arm, B-X, has to be counter-balanced by the moment caused by the abductors A that act on the shorter lever arm A-B. That is, the abductor musculature indicated by 'A' acts on a lever arm acting from the lateral aspect of the greater trochanter.

Table 3.2 indicates a summary of the range of force exerted on the hip for everyday activities.

Various actions imposed on the hip can now be explained. For the position of stance on one leg, the abductor musculature has to exert an equal moment in order to hold the pelvis level. The length of the lever arm of the body weight compared to that of the abductor musculature is about 2.5 to 1. In the stance phase of gait (see Chapter 4), the load on the femoral head is composed of the sum of the forces caused by the abductors and by the body weight. This load is about three times that of the body weight.

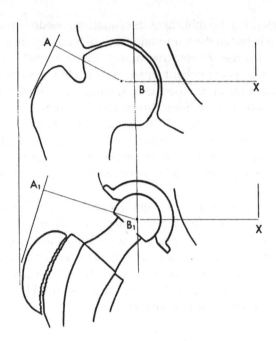

Figure 3.14   Biomechanical forces on hip (see text for discussion of symbols A, B, A_1, B_1, and X).
Source: Reproduced from Canale, S.T. and Beaty, J.H. (eds) (2007) *Campbell's Operative Orthopaedics,* 11 edn, Mosby Publishers, Philadelphia, PA.

Table 3.2   Maximum force at hip (expressed as a multiple of body weight).

Activity	Hip
Level walking slow	5
Level walking normal	5
Level walking fast	8
Upstairs	7
Downstairs	7
Up ramp	6
Down ramp	5

Source: After Paul, J.P. (1976) Loading on normal hip and knee joints and joint replacements, in *Advances in Hip and Knee Joint Technology* (eds M. Schaldach and D. Hommann), Springer-Verlag, Berlin, pp. 53–70.

For the action of raising the leg to the straight position, the load on the femoral head is also about three times that of the body weight.

For the condition of jumping, the load can be about ten times that of the body weight. Other actions such as ascending or descending stairs, moving on an incline, or arising from a chair, have the combined effect of increasing the torsional force acting on the hip joint.

In conclusion, increase in physical activity, and in body weight, causes a rise in the forces acting on the femoral component adjacent to the hip. The effect is loosening, bending, or breaking of the femoral stem.

3.6 Knee

The knee has to be able to provide mobility for the body and stability to accept forces due to body weight and the upward reaction of ground when the body undergoes movement and weight bearing.

Its position below the hip and above the ankle joints requires it to possess the anatomical characteristics needed for this twofold purpose. As discussed in Chapter 2, its function depends greatly on its articular cartilage (AC), synovial fluid and the meniscus. The AC possesses a low coefficient of friction. It is bathed in synovial fluid, which provides the viscosity needed to lubricate its sliding surface during flexion. The meniscus, sited between the tibia and femur, reduces the stress on the load-bearing surfaces of the knee by increasing the contact area between the femoral and tibial condyles. These features help it to deal with the forces arising from muscles acting on the hip and at the same time with those that surround the ankle.

The stresses imposed on the knee have given rise to its replacement, with the hip, being the most common type of joint replacement.

3.6.1 Anatomy of the Knee Joint

Figures 3.15 and 3.16 show the main parts of the knee. It is mainly composed of four bones.

The femur (thigh bone) is attached by ligaments to the tibia (shin bone) and encased at this articulation by the synovium-lined capsule. The fibula is sited below and parallel to the tibia. The patella (knee cap) rides on the knee joint as the knee bends. The knee joint is formed between the lower and upper regions of the femur and the tibia respectively. The lower end of the femur comprised two hemispherical structures, termed 'condyles': they are convex in shape, from side to side, and are accentuated to the rear of

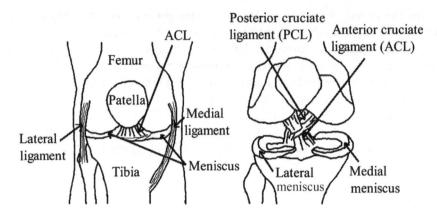

Figure 3.15   Anatomical features of the knee.
Source: After Moran, R. (2001) *The Mechanical Properties and Behavioural Characteristics of the Human Knee Joint Meniscus, PhD Thesis,* University of Edinburgh, Edinburgh.

the knee joint. The upper end of the tibia hollows out centrally, then flattens towards the periphery of the joint. The ends of the bones are covered with a cushion of hard, articular cartilage.

The knee is held together by two main ligaments that form a direct bond between the femur and the tibia, in the centre of the joint: the anterior (ACL) and posterior (PCL) cruciate ligaments are shown in Figures 3.15 and 3.16.

The ACL and PCL play a key role in stabilizing the function of the knee, and in tension exhibit considerable strength. In healthy knees they are effectively inelastic. Their length remains constant, on flexing and extension of the knee. These characteristics facilitate control of the rolling and gliding movements of the knee joint surfaces.

As described in Moran (2001), the PCL prevents the femur from sliding forward on the tibia (or the tibia from sliding backwards on the femur: the lateral view is shown in Figure 3.15).

The overall effect of these ligaments is to provide rotational stability to the knee. Indeed the knee has to ensure stability of the lower part of the body, whilst providing flexion and axial rotation. These anatomical features that affect stability may be clarified further from Figure 3.16.

The intrinsic stability of the knee is provided by the four primary ligaments, the capsules, the shape of the long structure and the menisci. The ligaments impose a tensile orientation limit on its motion. Such motion and

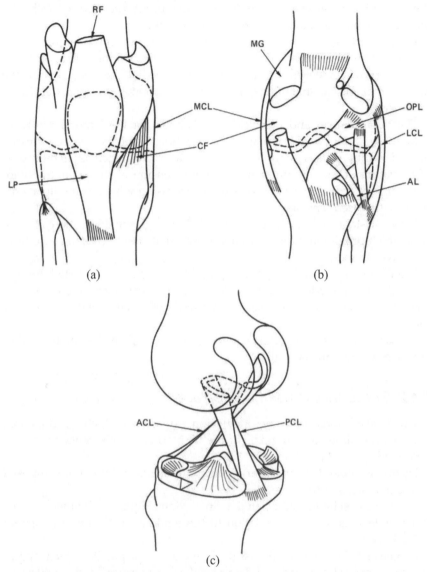

(a) (b)

(c)

Notation: RF: tendon; LP: ligamentum patellae; MCL: medial collateral ligament;
CF: capsular fibres; OPL: oblique popliteal ligament; LCL: lateral collateral ligament;
AL: arctuate ligament; ACL: anterior cruciate ligament; PCL: posterior cruciate ligament.

Figure 3.16 Detailed anatomical features of the knee: (a) Anterior (b) posterior aspects of capsule, (c) perspective view of cruciate ligaments and menisci. *Source:* After Bowker, P., Condie, D.N., Bader, D.L. and Pratt, D.J. (1993) *Biomechanical Basis of Orthotic Management*, Butterworth Heinemann, Oxford.

the mechanisms associated with the extension of the knee are controlled by its two articulating patellar and tibio-femoral joints.

Bony structures on the tibial plateau, termed 'intercondylar tubercles', interlock with the intercondylar notch of the distal femur. This 'interference' fit provides stability to the lateral and medial shear stresses on the knee. It remains capable of flexion and axial rotation, but hyperextension is inhibited.

The tibial plateau and menisci have channels for the femoral condyles, enhancing stability to flexion and axial rotation. Hyperextension is further limited by the effects of a couple arising from an anterior increase in the radius of curvature of the femoral condyles. On extension of the knee, the tibio-femoral joint is subjected to increased compression whilst in the ligaments there is increased tension.

Extrinsic stability to the knee is provided through the thigh and calf muscles, which provide the forces needed to articulate the knee. They also limit excessive stresses.

Extension is secured through the 'quadriceps mechanism' exerted through the 'patella'. The latter bone provides a pulley mechanism, increasing the efficiency of the effort of extension, and directing the action of the quadriceps forces.

These motions of the knee are complex; consequently its kinematic models are usually two dimensional.

3.6.2 Biomechanics of the Knee Joint

The normal motions of the knee are flexion, extension, abduction, adduction, and rotation about the long axis of the limb. These movements have been described in Chapter 2.

Flexion is dependent on the articular components and the limits imposed by the ligaments of the knee.

In the biomechanics associated with flexion, Figure 3.17 illustrates the extensor lever arm and its relation to the tendons and centre of rotation of the knee.

Canale and Beaty (2007) discuss the influence of the patella, which lengthens the extensor lever arm as it displaces the force vectors of the quadriceps and the tendons of the patella away from the centre of rotation of the knee.

The complexity of these movements of the knee emphasizes the significance of gait analysis in appraising the need for, and consequences of, knee replacements. From Figure 3.18, Laskin's (1991) interpretation of the kinematics of the knee joint can be discussed.

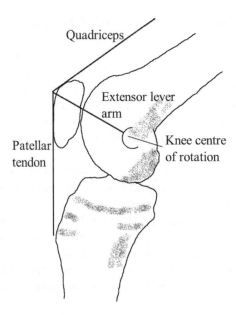

Figure 3.17   Biomechanics of knee flexion.
Source: Reproduced from Canale, S.T. and Beaty, J.H. (eds) (2007) *Campbell's Operative Orthopaedics,* 11 edn, Mosby Publishers, Philadelphia, PA.

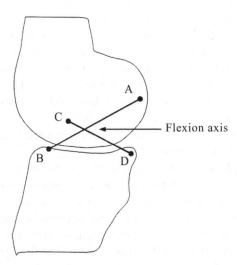

Figure 3.18   Biomechanical linkages of knee.
Source: Reproduced with permission from Laskin, R.S. (ed.) (1991) *Total Knee Replacement,* Springer-Verlag Publishers, London.

The line AB indicates the neutral fibre of the anterior cruciate ligament (ACL). Line CD represents the neutral fibre of the posterior cruciate ligament (PCL). The link between the tibial attachment sites is denoted by the line BD. The femoral insertion sites are represented by CA. The flexion axis occurs at the intersection of the links AB and CD. The two-dimensional motion of the knee joint is investigated for a case in which the femoral link is articulated about a fixed tibial link. In motion such as walking and running, the knee becomes almost at full extension, and undergoes mainly a rolling action.

The knee flexes on motions such as standing from a sitting position, or squatting. Sliding of the femur and tibia with respect to each other is the main action associated with this motion.

For the condition in which the tibia rotates on a stationary femur and, conversely, for the femur rotating on a stationary tibia, the point of intersection of the anterior and posterior (PCL) cruciate links is found to move in the posterior direction. For either motion, a plot of the centre of rotation, or 'centroid' is found to be a straight line for the tibial centroid; that for the femoral is elliptical.

The tibia and femur roll and slide as they move on each other. During flexion, the femur must slide forward on the tibia (solely rolling would cause the femur to fall off the back of the tibia, for flexing over 100°).

In the early phases of flexion, rolling is the main action. It is gradually replaced by sliding as flexion continues.

The mechanisms that give rise to axial rotation of the tibia with respect to the femur are caused by the ground reactive force, or 'passive rotation', the rotation applied by the musculature ('manual rotation') and the 'screw-home' ('automatic rotation'). This rotation is restrained by the ligaments and capsular structures. Passive restraint against, and the force causing, rotation derive from the musculature of the lower limb.

On extension of the knee, the automatic external rotation of the tibia on the femur arises (the 'screw-home' effect).

In the sagittal plane, the radii of curvature of the medial and lateral plateau differ. As the medial tibial plateau is concave superiorly, and that of the lateral convex, the articulating surfaces of the tibiofemoral joint are not congruent. From the distal position, the lateral femoral condyle is longer than that of the medial. The combined effects of unequal length, and of radii of curvature of the two femoral condyles, and the position of the four primary ligaments, cause the tibia to rotate externally with extension.

Laskin (1991) has summarized the findings of many researchers concerned with the forces on knee joints associated with loading.

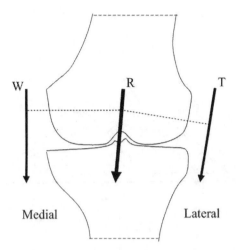

Figure 3.19   Effects of forces on knee (person standing on one leg).
Source: Reproduced with permission from Laskin, R.S. (ed.) (1991) *Total Knee Replacement,* Springer-Verlag Publishers, London.

For a person standing on both legs, each knee carries 43% of the body weight. When the person stands on one leg, the supporting knee has to support the partial body weight W suspended vertically from the centre of gravity of the body. Equilibrium is maintained by a force generated in the lateral tension band T. As shown in Figure 3.19, across the joint surface, the resultant force R is more than twice the body weight, and is located between the centres of curvature of the medial and lateral femoral condyles through the centre of the weight-bearing surfaces.

For a person walking, the compressive and shear forces at the knee joint can be as much as respectively six and two times the body weight. Even for a person walking on a level surface, the resultant forces across the knee joint are as high as seven times the body weight. The everyday actions of walking require that friction in the knee has to be low, especially with an average of 1.9 million steps taken per year (Laskin 1991).

3.7 Ankle

3.7.1 Anatomy of the Ankle Joint

The ankle joint is located at the junction of the foot and leg. Its bones are termed the 'tarsals'. Articular cartilage covers all the joint surfaces. Figure 3.20

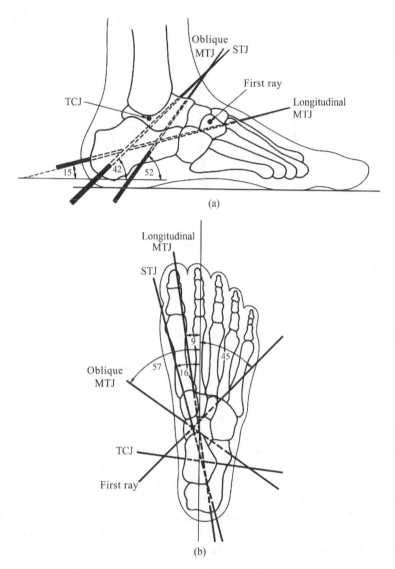

Notation: MTJ: mid-tarsal joint; STJ: subtalar joint; TCJ: talocrural joint; the first ray consists of the first metatarsal, the intermediate cuneiform, and the second metatarsal.

Figure 3.20   Ankle bones and joint axis: (a) sagittal plane; (b) transverse plane. *Source:* After Bowker, P., Condie, D.N., Bader, D.L. and Pratt, D.J. (1993) *Biomechanical Basis of Orthotic Management*, Butterworth Heinemann, Oxford.

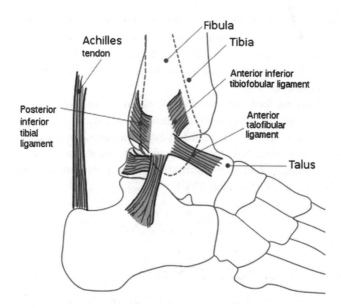

Figure 3.21 Ankle joint and ligaments.
Source: Adapted from Standring, S. (2008) *Gray's Anatomy: The Anatomical Basis of Clinical Practice, Expert Consult.* 40th edn, Churchill Livingstone Elsevier, London, U.K.

shows the positions of these bones of the ankle. The ankle joint connects the distal extremities of the tibia and fibula of the lower limb with the proximal end of the talus of the foot.

3.7.2 Biomechanics of the Ankle Joint

The joint is of the synovial hinge type. The joint is strongly supported by a set of ligaments (Figure 3.21). The stability that they impart enables the ankle to undergo flexion and extension termed respectively 'dorsiflexion' and 'plantar' flexion.

The axis of rotation of the ankle joint is noted to have a small tilt by 7° and 13°. This tilt causes the direction of movement of the foot to be misaligned with that of the proper sagittal plane, for plantar flexion and dorsiflexion.

Other joints associated with the foot are significant to the ankle.

Between the talus and calcaneus is situated the uniaxial subtalar joint. The means by which the foot provides mobility and the ankle and leg stability depends greatly on this joint – for example, how the foot and leg are positioned to carry the weight associated with exercise like running over uneven surfaces or around bends.

At the ankle and foot are also found the intertarsal and tarsometatarsal joints. They are nonaxial. These joints glide on each other. They enable movements that let the foot adjust its position over uneven surfaces, and can also let the foot and leg align for balance during weight-bearing activities. In walking the load on the ankle has been estimated to be approximately three to four times the body weight (Johnson 1981).

3.8 Foot

3.8.1 Anatomy of the Foot Joints

Figure 3.22 shows the main anatomical features of the foot. It possesses 20 bones, usually grouped into rear-, mid-, and forefoot. The calcaneus and talus bones make up the rear-foot. Between the talus and calcaneus is sited the subtalar joint.

The bones that comprise the mid-foot are the cuboid, medial intermediate and lateral cuneiforms and the navicular. The forefoot is composed of the metatarsals and phalanges of the toes.

3.8.2 Biomechanics of the Foot Joints

As the kinematic behaviour of the foot is closely linked with that of the ankle, the previous section has been used to include the basic biomechanics of the foot.

3.9 Toe

3.9.1 Anatomy of the Toe Joints

From the medial to lateral position, these digits comprise the (big) hallux, which is the most proximal (innermost), second (long), third (middle), fourth (ring), and fifth (little) toes. Their bones articulate around the metatarsals of

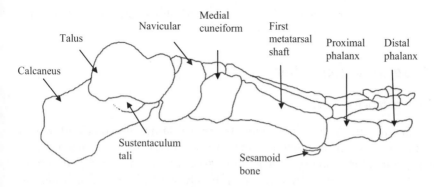

Medial view

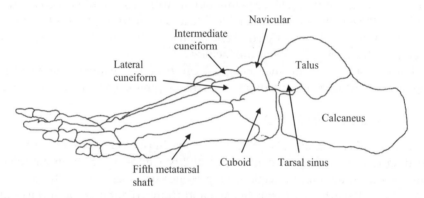

Lateral view

Figure 3.22   Anatomy of the foot and toes.
Source: Reproduced from Pratt, D., Tollafield, D., Johnson, G. and Peacock, C. (1993) Foot orthotheses, in *Biomechanical Basis of Orthotic Management* (eds P. Bowker, D.N. Condie, D.L. Bader and F.J. Pratt), Butterworth-Heinemann, Oxford, pp. 70–98.

the central part of the foot. Flexion is controlled by the tendons and muscles of the lower leg. The joints between the toe bones are called 'interphalangeal'.

3.9.2 Biomechanics of the Toe Joints

Toes provide balance, weight-bearing capacity and ability to provide thrust during movement like walking. Their biomechanical behaviour is closely

linked to that of the ankle and foot, as noted above, and is discussed by
Reimann and Marlovits (1992).

3.10 Degradation of Joints

3.10.1 Introduction

The term 'arthritis' is usually used to describe inflammation of a joint, which
is associated with pain, swelling and stiffness, often caused by degenerative
changes, infection or injury. At least 1 in 100 people suffer from the condition.
Arthritis can affect all joints in the body. Over 100 forms of arthritis are
known, with joint pain being a common symptom. The main types described
here are: osteoarthritis (OA), rheumatoid arthritis (RA), and post-traumatic
arthritis (the effects of which resemble OA, and arthritis due to sports injury).
The discussion will be restricted to one main joint, the knee, although in later
chapters effects on other joints are included.

With arthritis present, the knee joint can become unable to support the
upper body weight, and everyday and athletic physical activity is restricted.

3.10.2 Osteoarthritis (OA)

Osteoarthritis is a family of conditions, all sharing cartilage damage and loss,
leading to impaired movement and disability. Causes range from inherited
diseases to trauma and infection. Its presence becomes evident from wear
and tear of the articular cartilage (AC) and from irritation of the surfaces of
the tibia and patella bones. The articular cartilage and menisci that cushion
the knee joint from impact, and facilitate smoothness in its articulation, are
damaged by OA.

The degeneration begins with softening of the AC (termed 'fibrillation'),
which eventually wears away the surface of the articular cartilage. Without
this cartilage, direct grinding occurs at the bone surfaces of the tibiofemoral
and patellofemoral articulations. The bones erode, and bony spurs begin to
form, as indicated in Figure 3.23.

Minute shards of bone and cartilage start floating loosely in the space
between the joints. When the space becomes too narrow, the meniscus also
wears. The condition becomes marked by extreme pain.

3.10.3 Rheumatoid Arthritis (RA)

Rheumatoid arthritis is a chronic inflammatory disorder of the joint. The
body's immune system causes inflammation, which attacks the tissue and

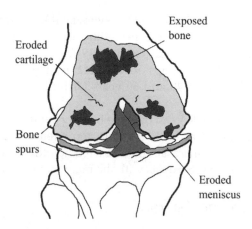

Figure 3.23 Effects of osteoarthritis on knee.
Source: Adapted from Oxford Knee Group (2010) *Your Knee Problem-Knee Arthritis*, Oxford, U.K. Available at: http://www.oxfordkneegroup.com/your-pathway/your-knee-problem/knee-arthritis/

bone and neighbouring ligaments. When the knee is affected by RA, its capsule is stretched by the inflammation. Even when the swelling decreases, the knee capsule does not maintain its original position. The joint can no longer be maintained in its proper location. The ligaments, tendons, and menisci become damaged. The cartilage can also be thinned by the effects of RA. The

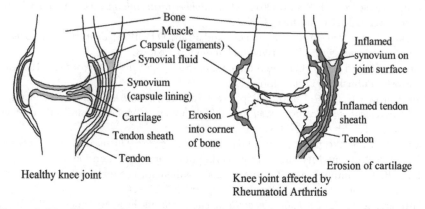

Figure 3.24 Effect of rheumatoid arthritis (RA) on healthy knee joint.
Source: After Arthritis Research UK (2011) *Rheumatoid Arthritis Information Booklet*, Arthritis Research U.K., Chesterfield, U.K.

disease can also affect the viscosity of the synovial fluids with adverse effects on the lubrication of the joint. These instabilities cause a rise in the mechanical stresses imposed on the meniscus (Gardner 1972, 1992).

Rheumatoid arthritis develops in about 1 in 100 people, most commonly those aged between 40 and 60 years (Arthritis Research UK Report 2011).

Figure 3.24 shows how RA can damage healthy joints.

When the pain due to arthritis becomes too intense and physical activity in any joint becomes increasingly restricted, joint replacement has to be considered. The next chapter deals with the main methods of assessment.

3.10.4 Infection and Trauma

Joint degradation can also stem from bone or wound infection, trauma, such as a car accident, or disorders arising from discontinuity in the blood supply to bone tissue and congenital defects (Court-Brown 2009). The consequence is often joint replacement.

References

Abrahams, P.H., Craven, J.L. and Lumley, J.S.P. (2011) *Illustrated Clinical Anatomy*, 2nd edn, Taylor & Francis Ltd, London.

Amis, A.A. (1998) Biomechanics of the elbow (Chapter 9), in *Joint Replacement in the Shoulder & Elbow* (ed W.A. Wallace), Butterworth-Heinemann, Oxford.

Arthritis Research UK Report (2011) *Rheumatoid Arthritis Information Booklet*, Arthritis Research U.K., Chesterfield, U.K.

Bowker, P., Condie, D.N., Bader, D.L. and Pratt, D.J. (1993) *Biomechanical Basis of Orthotic Management*, Butterworth Heinemann, Oxford.

Canale, S.T. and Beaty, J.H. (eds) (2007) *Campbell's Operative Orthopaedics*, 11th edn, Mosby Publishers, Philadelphia, PA.

Court-Brown, C. (2009) Femoral diaphyseal fractures, in *Skeletal Trauma: Basic Science, Management, and Reconstruction* (eds B.D. Browner, A.M. Levine and J.B. Jupiter), Elsevier Health Sciences, Oxford.

Dowson, D. and Wright, V. (eds) (1981) *An Introduction to the Bio-mechanics of Joints and Joint Replacement*, Mechanical Engineering Publication Ltd, London.

Ferris, B.D., Stanton, J. and Zamora, J. (2000) Kinematics of the wrist. Evidence for two types of movement. *J Bone Joint Surg Br* 82(2) 242–245.

Gardner, D.L. (1972) *The Pathology of Rheumatoid Arthritis*, Edward Arnold Publishers, Ltd, London.

Gardner, D.L. (1992) *Pathological Basis of the Connective Tissue Diseases*, Edward Arnold Publishers, Ltd, London.

Hamblen, D.L. and Simpson, A.H.R.W. (2010) *Adam's Outline of Orthopaedics*, 14th edn, Churchill Livingstone Elsevier, Edinburgh.

Hunsicker, P. (1955) *Arm Strength at Selected Degrees of Elbow Flexion*, technical report 45-548, Wright Patterson Air Force Base, Dayton, OH.

Johnson, G.R. (1981) Biomechanics of the upper limb, in *An Introduction to the Biomechanics of Joints and Joint Replacement* (eds D. Dowson and V. Wright), Mechanical Engineering Publication Ltd, London.

Laskin, R.S. (ed.) (1991) *Total Knee Replacement*, Springer-Verlag, London.

Lichtman, D.M. (ed.) (1988) *The Wrist and its Disorders*, Philadelphia, W.B. Saunders.

Lichtman, D.M., Schneider, J.R., Swafford, A.R. and Mack, G.R. (1981) Ulnar mid carpal instability-clinical and laboratory analysis. *Journal of Hand Surgery* **6** (5), 515–523.

McQueen, M.M. (1998) Redisplaced unstable fractures of the distal radius: a randomized prospective study of bridging versus non-bridging external fixation. *Journal of Bone and Joint Surgery – British Volume* **80-B** (4), 665–669.

McQueen, M.M., Hajducka, C. and Court-Brown, C. (1996) Redisplaced unstable fractures of the distal radius: a prospective randomised comparison of four methods of treatment. *Journal of Bone and Joint Surgery – British Volume* **78** (3), 404–409.

Moran, R. (2001) *The Mechanical Properties and Behavioural Characteristics of the Human Knee Joint Meniscus, PhD Thesis*, University of Edinburgh, Edinburgh.

Oxford Knee Group (2010) *Your Knee Problem-Knee Arthritis*, Oxford, U.K. Available at: http://www.oxfordkneegroup.com/your-pathway/your-knee-problem/knee-arthritis/

Paul, J.P. (1976) Loading on normal hip and knee joints and joint replacements, in *Advances in Hip and Knee Joint Technology* (eds M. Schaldach and D. Hommann), Springer-Verlag, Berlin, pp. 53–70.

Poppen, N.K. and Walker, P.S. (1978) Forces at the glenohumeral joint in abduction. *Clinical Orthopaedics and Related Research* 135, 165–170.

Pratt, D., Tollafield, D., Johnson, G. and Peacock, C. (1993) Foot orthotheses, in *Biomechanical Basis of Orthotic Management* (eds P. Bowker, D.N. Condie, D.L. Bader and F.J. Pratt), Butterworth-Heinemann, Oxford, pp. 70–98.

Reimann, R., and Marlovits, S. (1992) Biomechanics of the joints of the large toe. *Acta Anatomica* **144** (1), 30–35.

Souter, W.A. (1977) 'Total replacement arthroplasty of the elbow', in *'Joint Replacement of the Upper Limb: Conference'*. (eds Institution of Mechanical Engineers [IMechE]), Mechanical Engineering Publications for the Institution of Mechanical Engineers, London, U.K.

Standring, S. (2008) *Gray's Anatomy: The Anatomical Basis of Clinical Practice, Expert Consult*. 40th edn, Churchill Livingstone Elsevier, London, U.K.

Taleisnik, J. (1985) *The Wrist*, Churchill Livingstone, New York.

Tortora, G.J. and Derrickson, B.H. (2012) *Principles of Anatomy and Physiology*, John Wiley & Sons, Inc., New York.

University of California (2000) *Muscle Physiology*, www.muscle.ucsd.edu/musintro/ma.shtml (accessed 22 August 2012).

Unsworth, A. (1981) Cartilage and synovial fluid, in *Introduction to the biomechanics of joints and joint replacement*, (eds D. Dowson, and V. Wright), Mechanical Engineering Publications, London, pp. 486–510.

Wallace, W.A. (ed.) (1998) *Joint Replacement in the Shoulder and Elbow*, Butterworth-Heinemann, Oxford.

Youm, Y. and Yoon, Y.S. (1979) Analytical development in investigation of wrist kinematics. *Journal of Biomechanics* **12** (8), 613–621.

4

Methods of Inspection for Joint Replacements

4.1 Introduction

Prior to decisions on joint replacement, the condition of the bone joint and tissue structure has to be assessed, which involves gathering information on the background history of the patient. The clinical procedures performed are extensively documented, for example by Park and Hughes (1987). The sections below summarize the engineering principles of the methods used.

The interior of a diseased or damaged joint can be examined by arthroscopy. With this minimally invasive procedure, a small incision, typically about 4 mm long, is made at the joint. The arthroscope, a small fibre-optic camera, is inserted through the incision. The joint parts are irrigated with fluid to facilitate inspection; the camera can be connected to a video screen for ready examination. Arthroscopy can be performed on most joints that can be reached by the instrument. The most common joints examined by arthroscopy are knee, hip, shoulder, elbow, wrist, ankle and foot. For the knee it is used frequently in connection with damage to the meniscus, the anterior cruciate ligament and cartilage. With the hip, a wide range of conditions can be sought, including the presence of loose particles, infection, or osteoarthritis. Arthroscopy can also be employed to assess the condition of the joint prior to replacement.

The Engineering of Human Joint Replacements, First Edition. J.A. McGeough.
© 2013 John Wiley & Sons, Ltd. Published 2013 by John Wiley & Sons, Ltd.

4.2 Gait Analysis

Analysis proceeds by various techniques such as optical tracking by high-speed high resolution camera for example, as the patient walks over force plates fitted with strain gauges. An early pioneer of human motion photography was Eadweard Muybridge (1901). Data are collected on the weight, height of the patient and degree of possible flexion of the joints and muscles. Kinematic data may also be collected on the forces, moments and torques to which the joints are likely to be subjected. Figures 4.1 and 4.2 illustrate respectively a gait cycle, and typical forces measured during walking.

In the stance stage of walking the moments created about the hip, knee and ankle by the reaction of the ground forces are counterbalanced by the activities of the muscles. If the ground reaction force lies posterior to the knee, the knee flexion moment has to be opposed by the action of the knee extensors as shown in Figure 4.2.

For the standing (static) position the joints are neither flexing nor extending. In this case the ground reaction forces for both feet are equal in magnitude.

In other modes of gait, the stability of the knee joint does not necessarily need the activity of knee extensors. If the hip extensors contract, thereby causing a moment about the hip, there is a consequent posterior thrust at the knee. As a result, the knee is stabilized.

Whatling *et al.* (2011) have proposed a classification method that takes account of hip flexion, obliquity in the position of the pelvis, the frontal plane hip power and movement. This method can identify the improvements needed in hip biomechanics to restore gait to a normal state, in order to plan improvements in rehabilitation.

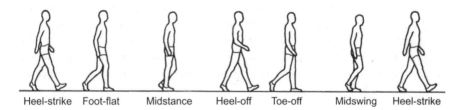

Heel-strike   Foot-flat   Midstance   Heel-off   Toe-off   Midswing   Heel-strike

Figure 4.1 Gait cycle for right leg.
Source: Reproduced from Bowker, P., Condie, D.N., Bader, D.L. and Pratt, D.J. (1993) *Biomechanical Basis of Orthotic Management,* Butterworth Heinemann, Oxford.

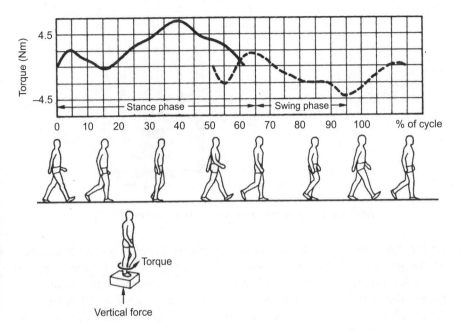

Figure 4.2 Forces and moments during walking for gait cycle.
Source: Reproduced from Bowker, P., Condie, D.N., Bader, D.L. and Pratt, D.J. (1993) *Biomechanical Basis of Orthotic Management,* Butterworth Heinemann, Oxford.

Sandholm *et al.* (2011) have used gait analysis in order to analyse kinematic models of knee joints, using a novel inverse dynamics approach. Inverse dynamics is based on investigation of generalized forces acting on joints for a specific motion. Information is obtained from data such as body mass, inertia, and kinematics in order to solve the second law of motion. They evaluate the moments on the hip joint for conditions of flexion, adduction and rotation. Similarly for the knee, the moments are calculated for flexion, adduction and rotation.

4.3 X-ray

X-rays are a type of electromagnetic radiation, of relatively high energy and short wavelength. As the X-rays are generated in all directions, lead shields are used to absorb them in undesired regions. For the X-rays impacting

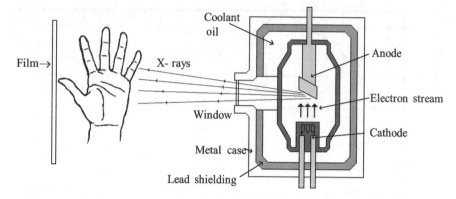

Figure 4.3 X-ray examination of human hand joints.

on the desired target, the amount of absorption depends on variables such as energy of the radiation, density of the material struck by the rays, for example, articular tissue and its properties. Following its attenuation as it passes through the material, the X-ray strikes the sensitized plate or film. A two-dimensional image is produced of the specimen through which the X-ray has been transmitted.

An example of the function of an X-ray machine for the examination of joints is shown in Figure 4.3.

Interpretation of the X-ray depends on the characteristics of the target. The denser the target the higher the attenuation and the whiter (or more blank) becomes the image on the X-ray film. As an example, a metal implant is recorded as deeply white. The attenuation with bone is less than that of metal. Therefore its X-ray image appears less white than that of the latter. Soft tissues are shown as grey images, depending on their density. For instance, as there is little attenuation in bone it will be imaged as a dark grey area between the ends of the bone.

Inspection of the X-ray image should reveal irregularities such as variation in density of soft tissue or bone, or regions where bone density has been reduced owing to osteoporosis. At the joints, the effects of arthritis should be evident by reduction in the width between the articulating bone surfaces, caused by loss of articular cartilage. An example of an X-ray of a knee joint is shown in Figure 4.4. Further examples of X-rays that demonstrate the narrowing of the width between the bone surfaces due to loss of articular cartilage are available in Solomon *et al.* (2005).

After joint replacement surgery the X-ray can be used to inspect the location of the prosthesis, as illustrated in Figure 4.5.

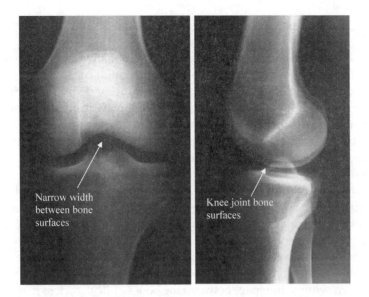

Figure 4.4 X-ray of knee joint.
Source: Reproduced with permission from Liow, R.Y.L., McNicholas, M.J., Keating, J.F. and Nutton, R.W. (2003) Ligament repair and reconstruction in traumatic dislocation of the knee. *Journal of Bone and Joint Surgery* **85-B** (6), 845–851.

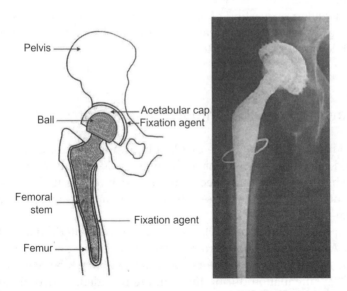

Figure 4.5 (a) X-ray of hip replacement with (b) location of prosthesis.
Source: Image courtesy of K. Skalski.

More detailed inspection of soft tissue can be obtained by use of 'contrast media'. The injection of a nontoxic radiopaque liquid, constituents of which possess a high atomic number, facilitates a more accurate profile of the joint. Fluoroscopic methods are sometimes used in conjunction with contrast media in order to procure further improvement in detail. The fluoroscopic image is viewed on a monitor, coupled to an image intensifier. The digitization and computer storage of the image are then undertaken.

As X-rays provide two-dimensional representation, two views at right angles are usually taken. The X-rays are used to examine the condition of soft tissue, bone and joints. Swelling or wasting of soft tissue is evaluated. Deformity or irregularity in the bone density is detected. Reduction in bone density or destruction of the cortices may be an indication of osteoporosis, increased density, a sign of sclerosis (stiffening of the local bone tissue). The X-ray examination of the joints will reveal the profile of the articulating bones. Narrowing of the space between the bones indicates a loss in thickness of (radiolucent) articular cartilage. The bone ends are examined. Evidence of bone flattening, erosion, cavitation or sclerosis indicates the presence of arthritis. At the margins of the joints the appearance of osteophytes is a sign of osteoarthritis; erosion signifies inflammatory disorders such as rheumatoid arthritis (RA).

The soft tissues, bone and joints are examined in sequence. Characteristics such as swelling or wasting of soft tissue are checked. Bone deformity or irregularity is next sought. Areas of destruction in the cortices or areas of reduced density are a sign of osteoporosis.

4.4 Tomography and Computed Tomography (CT)

Neighbouring tissue may cause overlapping X-ray images. Clarity may be improved by tomography. In this technique, the X-ray tube and its film are moved simultaneously in a specific orbit, and in opposite directions. The movement is such that the cross-sectional X-ray image is more defined in a particular plane.

In computer tomography (CT), also known as computerized axial tomography (CAT), the X-ray tube is rotated usually over 360° around the patient for about 4–5 s. The X-ray detectors may rotate with the tube. Measurements of the X-ray attenuation during the scan are recorded. A two-dimensional cross-sectional outline of the bone and soft tissue is obtained, data adapted

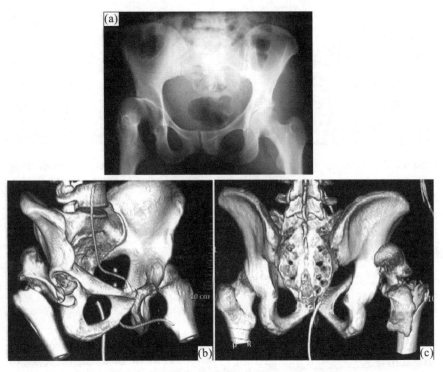

Figure 4.6   (a) Plain X-ray of hip dislocation, (b) and (c) 3D CT image of a hip dislocation.
Source: Reproduced from Rodriguez-Martin, J., Prettel-Mazzini, J., Porras-Moreno, M.A. *et al.* (2010) A polytrauma patient with an unusual posterior fracture-dislocation of the femoral head: a case report. *Strategies in Trauma and Limb Reconstruction* **5**, 47–51.

to provide a three-dimensional image. The thinner the thickness of the slice the greater is the detail obtained (although the radiation may be more intense).

Figure 4.6 shows how a hip dislocation image in a plain X-ray is clearer than that obtained by a three-dimensional CT scan. Figure 4.7 illustrates the detailed bone contour information which can be obtained from CT scans. A major disadvantage of CT scans compared to X-rays is that the patient is exposed to higher radiation levels.

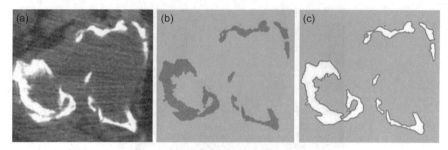

Figure 4.7   Bone contour detection in CT scan. (a) CT image, (b) bone tissue of the specified radiological density, (c) contours of detected areas.
Source: Reproduced from Skalski, K., Kwiatkowski, K., Domanski, J. and Sowinski, T. (2006) Computer-aided reconstruction of hip joint in revision arthroplasty. *Journal of Orthopedics and Traumatology* **7**, 72–79.

4.5 Radionuclide Scanning

X-rays may not reveal the extent of stress fractures in bone, the presence of which may give rise to the need for joint replacement, or to the presence of infection or tumours in bone. In these cases, radionuclide scanning may be used. Following the intravenous injection of a radioisotope compound, such as fluorine-18, into the tissue, a gamma camera (or rectilinear scanner) is used to locate the presence of the isotope, and consequent activity in the tissue, including production of new bone.

4.6 Ultrasonography

Ultrasound scans are radiation free. They provide a useful way of investigating soft tissue damage to joints. A transducer is used to generate high frequency sound waves, which are directed to enter tissue to a depth of several centimetres. Reflection, at the interface between tissue, of some of the waves, back to the transducer gives rise to electrical signals and then to images, shown on a monitor screen or plate. Various shades of grey on the

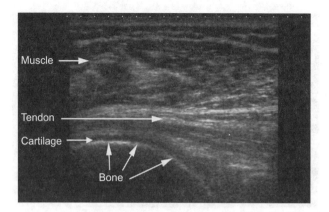

Figure 4.8   Ultrasound scan of a shoulder joint.
Source: Reproduced by permission from Mullaney, P. (2013) *Ultrasound,* London Sports Orthopaedics. http://www.sportsortho.co.uk/left navigation/ imaging/ultrasound (accessed 2 July 2013).

plate or monitor yield an outline of the anatomy, or dynamic image. Figure 4.8 shows the results from an ultrasound scan of a shoulder joint.

4.7 Magnetic Resonance Imaging (MRI)

Magnetic resonance imaging is effective in investigating bone tumours, injury to cartilage and ligaments in all joints, commonly meniscal tears and cruciate ligament injuries in the knee. The technique does not use ionizing radiation as needed for X-rays or CT scans.

In MRI, tissue is subjected to a static magnetic field. Radio frequency emission arises from the atoms and molecules in the tissue. Those that contain substantial amounts of hydrogen, such as cancellous bone, will yield signals of high intensity and bright images. Tissue with less hydrogen – typically cortical bone, ligament and tendons – produce dark (almost black) images.

Cartilages, muscle and spinal canal tissue, which contain medium amounts of hydrogen, produce grey images. As is evident from comparison between Figures 4.9 and 4.10, very clear images of the various tissues and organs can be procured by control of the MRI process variables, such as the magnetic field strength.

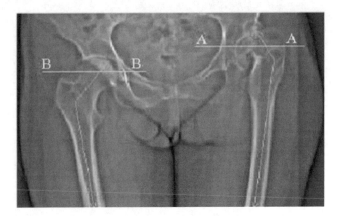

Figure 4.9 X-ray of hip joints illustrating deformity.
Source: Image courtesy of K. Skalski.

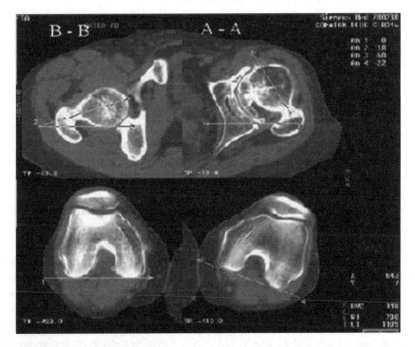

Figure 4.10 Magnetic resonance imaging scan of same hip joints illustrating clearer images compared to X-rays.
Source: Image courtesy of K. Skalski.

References

Liow, R.Y.L., McNicholas, M.J., Keating, J.F. and Nutton, R.W. (2003) Ligament repair and reconstruction in traumatic dislocation of the knee. *Journal of Bone and Joint Surgery* **85-B** (6), 845–851.

Mullaney, P. (2013) *Ultrasound*, London Sports Orthopaedics. http://www .sportsortho.co.uk/left navigation/imaging/ultrasound (accessed 2 July 2013).

Muybridge, E. (1901) *The Human Figure in Motion: An Electrophotographic Investigation of Consecutive Phases of Muscular Actions*, Chapman & Hall., London.

Park, W.M. and Hughes, S.P.F. (1987) *Orthopaedic Radiology*, Black Scientific Publications, Oxford.

Rodriguez-Martin, J., Prettel-Mazzini, J., Porras-Moreno, M.A. *et al.* (2010) A polytrauma patient with an unusual posterior fracture-dislocation of the femoral head: a case report. *Strategies in Trauma and Limb Reconstruction* **5**, 47–51.

Sandholm, A., Schwartz, C., Pronost, N. *et al.* (2011) Evaluation of a geometry-based knee joint compared to a planar knee joint. *Journal of Image and Vision Computing* **27**, 161–171.

Solomon, L., Warwick, D.J. and Nayagam, S. (2005) *Apley's Concise System of Orthopaedics and Fractures*, 3rd edn, CRC Press, Boca Raton, FL, U.S.A.

Whatling, G., Holt, C. and Beynon, J.J. (2011) Hip gait improvement analysis – patient specific targeted improvement, I.MechE. event proceedings: Engineers and Surgeons: Joined at the Hip, 1–3 November, pp. 25–28.

5

Materials in Human Joint Replacement

5.1 Introduction

Joints in the body may be replaced by engineering materials such as metal alloys, ceramics and polymers. Metals have the mechanical strength and stiffness necessary for load bearing. Ceramics – compounds of metallic and nonmetallic elements – and especially the oxide ceramics, which are wear-resistant, along with calcium phosphate and bioactive glass, exhibit attractive osteoconductivity (that is, encouragement of natural bone growth). These implanted materials affect the local mechanical and chemical region of the body in which they are sited. They can also influence the characteristics of the entire body. Polymers-on-metal prostheses exhibit lower frictional effects than their all-metal counterparts and any debris from their wear is considered to have less deleterious biological effects.

5.2 Alloy Metals

Many metals including iron, chromium, cobalt, nickel, titanium, and molybdenum are incorporated into metallic alloys for implants, and can be accepted by the body albeit in minute quantities. The first alloy used was vanadium steel for the manufacture of plates and screws, later replaced by other metals, owing to its unanticipated low resistance to corrosion. An alternative was stainless steel.

The Engineering of Human Joint Replacements, First Edition. J.A. McGeough.
© 2013 John Wiley & Sons, Ltd. Published 2013 by John Wiley & Sons, Ltd.

Stainless steel, type 18-8, and in particular grade 316 containing molybdenum, is stronger and more resistant to corrosion than vanadium.

Most metals normally used in joint replacement are either iron-, cobalt-, or titanium- based alloys. These alloys clearly vary in both chemical composition and microstructure. The microstructure achieved is dependent on the manufacturing processes adopted. Production methods, such as casting and forging, are discussed in detail in Chapter 6.

The characteristics of the alloy play a key part in their use as prostheses. Their mechanical and corrosion-resistance properties stem from their chemical constituents and the way that their microstructure is produced.

The manufacturing process also influences the amount of defects (such as voids, slag) in the alloy and the porosity of its surface. Inclusions weaken the implant. Surface defects can act as stress risers and may cause crevice corrosion.

The alloys are also subject to metallurgical attack, primarily by corrosion, caused by the synovial fluids in the body. To overcome the possibility of corrosion, alloys for prostheses are chosen that exhibit a passive oxide layer on their surface, which inhibits the effect of corrosion. Passivation, or control of the thickness of the passive layer, can be achieved by electrochemical techniques, including polishing and etching.

The alloys should be biocompatible. That is, their presence in the body should not lead to an adverse or allergic reaction that, for example, might cause bone resorption and possible loosening of the implant.

The strength and stiffness of alloy metals that make them suitable for load bearing in many engineering applications, also make them attractive in joint replacement. In the body, however, the risk of their corrosion means that generally only three alloy metals that are sufficiently resistant to corrosion are employed: stainless steel, cobalt-chromium and titanium alloys.

5.2.1 Stainless Steel

These conditions for implanted materials may be achieved from austenitic stainless steel. It contains few impurities; its passive surface renders it corrosion resistant, and it is largely biocompatible. Stainless steel is composed of three main elements – iron, nickel and chromium. Chromium in the alloy promotes the formation of the oxide surface layer, which inhibits corrosion.

Stainless steel implants can either be forged or cast. As discussed later, the pioneering orthopaedic surgeon, Charnley, chose stainless steel as his hip replacement metal. Owing to fracture of the neck of the femoral component, the alloy was later replaced with cobalt and titanium alloys. Although it is

Table 5.1 Typical percentage by weight constituents of 316 and 316L stainless steel.

Element	316	316L
Carbon	0.08 max	0.03 max
Manganese	2.00 max	2.00 max
Phosphorus	0.03 max	0.03 max
Sulphur	0.03 max	0.03 max
Silicon	0.75 max	0.75 max
Chromium	17.00–20.00	17.00–20.00
Nickel	12.00–14.00	12.00–14.00
Molybdenum	2.00–4.00	2.00–4.00

Source: Adapted from American Society for Testing and Materials Standards (1980) *Annual Book of ASTM Standards*, American Society for Testing and Materials Standards, Philadelphia, PA.

still used sometimes for implants in elderly patients, whose life expectancy and physical activity are limited, instead of more expensive alternatives.

Austenitic stainless steel types 316 and 316L are commonly used for implants. They are hardened by cold working, are nonmagnetic and exhibit useful resistance to corrosion. The presence of molybdenum and chromium enhances resistance to corrosion. Their main constituents are given in Table 5.1.

The main mechanical properties of 316 and 316L stainless steels are shown in Table 5.2.

Table 5.2 Main mechanical properties of stainless steel (types 316 and 316L).

Condition	Ultimate tensile strength, min (MPa)	Yield strength (0.2% offset), min (MPa)	Elongation (50.8 mm bar length), min, %	Rockwell hardness, max
	(type 316)			
Annealed	(515)	(205)	40	95 HRB
Cold finished	(620)	(310)	35	–
Cold worked	(860)	(690)	12	300–350
	(type 316L)			
Annealed	(505)	(195)	40	95 HRB
Cold finished	(605)	(295)	35	–
Cold worked	(860)	(690)	12	–

Source: Adapted from American Society for Testing and Materials Standards (1980) *Annual Book of ASTM Standards*, American Society for Testing and Materials Standards, Philadelphia, PA.

The manufacture of implants from stainless steels is complex owing to variations in the properties caused by heat treatment or cold working. For example, these stainless steels rapidly work-harden. Cold working cannot be used without intermediate heat treatment and the latter should not promote the production in the grain boundaries of chromium carbide, the presence of which can lead to corrosion. Likewise, their welding is undesirable. Heat treatment can cause distortion of the implant, although it can be controlled. Oxide scales can arise on the surface after heat treatment and have to be removed by either chemical acid etching methods or mechanical sandblasting. On removal of the scales, the implant is given a polished or matt finish (these qualities are discussed in Chapter 6). It is then cleaned, degreased, and given a protective film by passivation in nitric acid. Finally it is washed and cleaned, and sterilized, then sealed.

Despite its attractions, when stainless steel was first used for implants, some corrosion was observed, albeit at a slow rate. The alloy is now employed when temporary implants are needed, and has been superseded by the alloys described below, which have superior corrosion resistance.

5.2.2 Cobalt-Based Alloys

Cast cobalt-chromium-molybdenum alloys have generally replaced stainless steel for implants, having a superior fatigue life. They exhibit considerable wear and corrosion resistance and satisfactory biocompatibility. Metallurgical drawbacks associated with the casting process such as large grain size and porosity have led to their manufacture by alternative techniques including forging and hot isostatic pressing (HIP), which reduce the extent of these adverse features.

The compositions of some cobalt-based alloys employed in implants are summarized in Table 5.3.

A solid solution is obtained up to 65 weight per cobalt, the remainder being chromium. Finer grains are obtained through the addition of molybdenum, which promotes higher strength after casting and forging. Cold working is used to increase the strength of Co-Ni-Cr-Mo alloys, although the cold working of large implants such as hip joint stems can prove difficult. Hot forging of Co-Ni-Cr-Mo alloy implants is commonly used.

The inadequate frictional properties of wrought cobalt-nickel-chromium-molybdenum alloys mean that it is unsuitable for the bearing surfaces of joint implants. On the other hand it is suitable for the stems of hip joint replacements, for which long service life is needed without the onset of fracture or stress fatigue. Table 5.4 shows the principal mechanical

Table 5.3 Constituents by percentage weight of cobalt-based alloys.

Element	Cobalt-chromium-molybdenum		Cobalt-chromium-tungsten-nickel		Cobalt-nickel-chromium-molybdenum	
	Min.	Max.	Min.	Max.	Min.	Max.
Cr	27.0	30.0	19.0	21.0	19.0	21.0
Mo	5.0	7.0	–	–	9.0	10.5
Ni	–	2.5	9.0	11.0	33.0	37.0
Fe	–	0.75	–	3.0	–	1.0
C	–	0.35	0.05	0.15	–	0.025
Si	–	1.00	–	1.00	–	0.15
Mn	–	1.00	–	2.00	–	0.15
W	–	–	14.0	16.0	–	–
P	–	–	–	–	–	0.015
S	–	–	–	–	–	0.010
Ti	–	–	–	–	–	1.0
Co	Balance	Balance	Balance	Balance	Balance	Balance

Source: After American Society for Testing and Materials Standards (1980), *Annual Book of ASTM Standards*, American Society for Testing and Materials Standards, Philadelphia, PA.

properties of these alloys. Their moduli of elasticity range from about 220 to 234 GPa, which is higher than stainless steel and may affect the load transfer to bone.

Implants made from cobalt-chromium-molybdenum alloy are manufactured by lost wax or investment casting. Coarser grains are obtained with higher mould temperatures in the lost-wax process, at which temperatures larger carbide precipitates may also arise and subsequently less brittle materials.

Table 5.4 Mechanical properties of cobalt-based alloys.

Property	Cast CoCrMo	Wrought CoCrWNi	Solution annealed	Cold worked and aged	Fully annealed
Tensile strength (MPa)	655	860	795–1000	1790	600
Yield strength (0.2% offset) (MPa)	450	310	240–655	1585	276
Elongation	8	10	50.0	8.0	50
Reduction of Area (%)	8	–	65.0	35.0	65
Fatigue strength (MPa)	310	–	–	–	340

Source: After American Society for Testing and Materials Standards (1980) *Annual Book of ASTM Standards*, American Society for Testing and Materials Standards, Philadelphia, PA.

Table 5.5 Constituents of unalloyed (Grades 1 to 4) and alloyed (Ti6Al4V) titanium.

Element	Grade 1	Grade 2	Grade 3	Grade 4	Ti6Al4V[a]
Nitrogen	0.03[b]	0.03	0.05	0.05	0.05
Carbon	0.10	0.10	0.10	0.10	0.08
Hydrogen	0.015	0.015	0.015	0.015	0.0125
Iron	0.20	0.30	0.30	0.50	0.25
Oxygen	0.18	0.25	0.35	0.40	0.13
Titanium			Balance		

Notes:
[a] aluminium 6.00 wt% (5.50–6.50), vanadium 4.00 wt% (3.50–4.50), and other elements 0.1 wt% maximum or 0.4 wt% total.
[b] All are maximum allowable weight percentage.
Source: Park, J.B. and Lakes, R.S. (1992) *Biomaterials: An Introduction*, 2nd edn, Plenum Press, New York.

5.2.3 Titanium-Based Alloys

The use of titanium for implants began in the 1930s. Its attraction was low density – 4500 kg/m^3, c.f. 316 stainless steel, 7900 kg/m^3. Its alloy Ti6Al4V, the chemical composition of which is shown in Table 5.5 is generally used for implants (the other four grades of unalloyed titanium recommended for implants are also listed).

Its composition may be summarized as follows. It is allotropic, and has the hexagonal close-packed form up to 882 °C (α-Ti), above which it has a body-centred cubic form. Two key alloying elements are aluminium and vanadium. Their aluminium element means that they have desirable strength and resistance to oxidization at high temperatures (300–600 °C). Owing to their single phase they cannot be strengthened by heat treatment.

Table 5.6 shows the mechanical properties of commercially pure and 6Al4V titanium alloy: the higher the impurity the greater the strength and lower the ductility.

Table 5.6 Mechanical properties of unalloyed (Grades 1 to 4) and alloyed (Ti6Al4V) titanium.

Properties	Grade 1	Grade 2	Grade 3	Grade 4	Ti6Al4V
Tensile strength (MPa)	240	345	450	550	860
Yield strength (0.2% offset) (MPa)	170	275	380	485	795
Elongation (%)	24	20	18	15	10
Reduction of area (%)	30	30	30	25	25

Source: After Park, J.B. and Lakes, R.S. (1992) *Biomaterials: An Introduction*, 2nd edn, Plenum Press, New York.

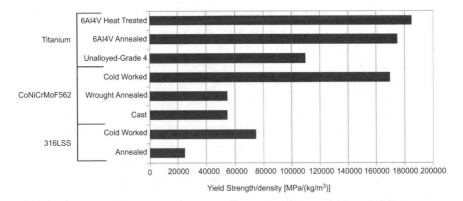

Figure 5.1 Comparison of yield strength per density for joint replacement alloys (F562 refers to ASTM notation).
Source: Modified from Park, J.B. and Lakes, R.S. (1992) *Biomaterials: An Introduction*, 2nd edn, Plenum Press, New York, Figure 5-9, p. 92, after Hille (1966). With permission, from the *Journal of Materials* (1966), copyright ASTM International, 100 Barr Harbor Drive, West Conshohocken, PA 19428.

Its strength is noted to be lower than that of 316 stainless steel or the cobalt-based alloys, and is comparable with that of annealed 316 stainless steel or cast Co-Cr-Mo alloy (see Figure 5.1). The strength per density of titanium alloy is greater than that of other implant metals. However owing to its low shear strength, it is unsuitable for use in bone screws and plates. The formation of a solid TiO_2 oxide surface layer passivates the alloy, rendering implants almost impervious to corrosion in the body.

As titanium is reactive at high temperatures, burning readily in oxygen, an inert atmosphere is used for high-temperature processing. Alternatively vacuum melting is used. As oxygen diffuses in titanium, and the alloy is embrittled by dissolved with oxygen, hot working or forging is performed at temperatures below 925 °C.

The alternative – machining at room temperatures – is rendered difficult by titanium galling (adhesive wear) or seizing of cutting tools. In order to overcome these drawbacks, very sharp tools running at slow speeds and high feed rates are used. Unconventional machining techniques such as electrochemical machining (ECM) are an alternative, although electrolyte solution choice renders difficult even these nonmechanical methods of machining.

Titanium-aluminium-vanadium alloy (Ti6Al4V) is a common choice for joint replacement. Its modulus of elasticity is about one-half of that of stainless steel and cobalt-chromium-molybdenum. Its modulus of elasticity facilitates the transfer of load to bone, although its stiffness is about five times greater than that of cortical bone.

The fatigue strength of titanium and cobalt alloys are similar. However the fatigue strength of this titanium alloy is more readily reduced by the presence of surface defects or other stress-raising surface features. Coatings of commercially pure titanium, bead or matt, on the surface are aimed at encouraging bone growth. Ti6Al4V bonding with bone is considered to be superior to that of cobalt alloy owing to its low wear resistance (and high coefficient of friction). Nitriding and nitrogen ion implantation have both been shown to increase the surface hardness and wear resistance. The latter technology has already been adopted in total knee replacements for titanium femoral components.

5.2.4 Tantalum Trabecular Metal

Tantalum possesses attractive qualities of corrosion resistance and biocompatibility. Vapour deposition methods can be used to manufacture a trabecular form of this element. This structure is 80% porous. Its compressive strength and elastic modulus are similar to that of bone. These characteristics are claimed to allow bone ingrowth that can be two to three times higher than that obtained with other porous coatings, including a strong attachment to the parent trabecular metal implant.

5.2.5 Magnesium Alloys

Although the alloys discussed above remain the principal metals used in joint replacement, magnesium alloy is receiving increasing attention. One attraction is its low density (1740 kg/m^3); (c.f. titanium 4500 kg/m^3). Amongst its many disadvantages are difficulty in casting, low strength and resistance to creep, fatigue and wear. It is also prone to corrosion. These drawbacks would seem to eliminate magnesium alloy as a material for joint replacement, yet its other mechanical properties have prompted recent interest. It has a lower elastic modulus than titanium (45 GPa c.f. 107 GPa respectively), and is closer to that of human cortical bone (15 GPa). It has the potential, therefore, to reduce the effects of stress shielding. It has already been employed as a lightweight resorbable material for bone repair: see for example Staiger *et al.* (2006), and Li *et al.* (2008).

5.3 Ceramics

5.3.1 Structure

Ceramics are made from inorganic materials with nonmetallic properties. General engineering examples are magnesium oxide, MgO, and silicon carbide, SiC, which both exhibit characteristic cubic crystal structures, albeit of different types. Their general structure may be presented in the form A_mX_n, where A and X represent respectively a metal and a nonmetallic element, and where m and n are integers. When the ions that make up the structure are of similar size, a simple body centred cubic (BCC) structure is obtained. If their sizes are different a face-centred cubic (FCC) structure arises.

In joint replacements, aluminium and chromium oxides are widely used. Their structure is of the A_2X_3 type, with a hexagonal close packing (HCP) form. Ceramics are generally produced in the following stages. Firstly the constituent materials are mixed into slurry (a water-based medium), or into a clay, which is a solid plastic mass containing water, then formed into the shape required, for example by moulding the casting. Next they are dried, normally in air, and fired in a furnace to expel any residual water and procure a desired grain structure and final dimensions. After firing, a final surface polish can be performed. Occasionally a glass-like finish is required, achieved by adding a 'glaze' substance to the slurry, the piece being refired at a lower surface temperature.

5.3.2 Mechanical Properties

Ceramics are attractive for joint replacement owing to their suitable compressive strength and resistance to abrasion. However, they exhibit brittleness. Typical mechanical properties of ceramic, that is non-resorbable (unaffected by the chemical effects of joint replacement implantation) are included in Tables 5.7 and 5.8.

Aluminium oxide is used widely in total hip replacements owing to its hardness – 19.6 to 29.9 GPa – and characteristic low friction. Carbon-silicon carbide and carbon-carbon fibre composites can also be produced in bulk form and then machined to the form required. Their wear rates are considerably less than that of UHMWPE (see section 5.4). Consequently they may be considered for use in the replacement of metal parts in joint replacements.

The mineral component of bone contains crystalline calcium phosphate, the characteristics of which resemble hydroxyapatite, that is $Ca_{10}(PO_4)_6(OH)_2$.

Table 5.7 Typical mechanical properties of ceramics.

Material	Tensile modulus (GPa)	Compressive strength (MPa)	Flexural strength (MPa)
'Commercial' carbon	12		90
Isotropic carbon	25	400	150
Vitreous carbon	24		170
Carbon-silicon carbide Composite (50: 50)	100	1000	220
Carbon-carbon fibre Composite (30: 70)	140	800	800
Aluminum oxide (>99.5%)	380	4000	400

Source: Adapted from Black, J. (1988) *Orthopaedic Biomaterials in Research and Practice,* Churchill Livingstone, New York.

Indeed synthetic calcium phosphate has been manufactured for application as artificial bone.

Mechanical and other properties for calcium phosphate are given in Table 5.9.

Precipitation of hydroxyapatite (HA) can be produced from aqueous solutions of $Ca(NO_3)_2$ and $Na-H_2-PO_4$. HA is obtained in a powder crystallization form by calcination for three hours at 800 °C followed by filtering and drying. The HA is then pressed into the form required and sintered at 1050 to 1200 °C for three hours (Park and Lakes, 1992).

In bone, HA is osteoconductive and is amenable to the presence of collagen and soft tissue. Ceramic implants to which tissue may attach are therefore attractive in joint replacements.

Surface active glass ceramics, such as polycrystalline ceramics, produced by nucleation and growth of crystals less than one μm in diameter, in which 10^{12} to 10^{15} nuclei per cm^3 are utilized, are used in direct bonding with bone. An example is $SiO_2-CaO-Na_2O-P_2O_5$. The bonding with bone relies on

Table 5.8 Mechanical properties of Alumina (Al_2O_3).

Flexural strength	>400 MPa
Elastic modulus	380 GPa
Density (kg/m^3)	3800–3900

Source: From American Society for Testing and Materials Standards (1980) *Annual Book of ASTM Standards,* American Society for Testing and Materials Standards, Philadelphia, PA.

Table 5.9   Properties of calcium phosphate.

Elastic modulus (GPa)	40–117
Compressive strength (MPa)	294
Bending strength (MPa)	147
Hardness (Vickers, GPa)	3.43
Poisson's ratio	0.27
Density (theoretical kg/m^3)	3160

Source: After Park, J.B. and Lakes, R.S. (1992) *Biomaterials: An Introduction*, 2nd edn, Plenum Press, New York.

the simultaneous formation of a layer of calcium phosphate and SiO_2-rich substance on the surface. Glass-ceramics have a low coefficient of thermal expansion, for example 10^{-7} to 10^{-5} per degree Celsius. Their tensile strength ranges from about 83 to 200 MPa. They are resistant to scratching and abrasion but are brittle materials. They can be used for fillers for bone cement and as coating materials, but not for load-bearing use in joint replacements.

A ceramic that can fill defects in bone or enhance healing and that can be gradually resorbed and replaced by tissue has long been sought. Recent developments include calcium-phosphorous ceramics.

Titanium oxide (TiO_2) has been assessed for use in bone cement. Improved prosthesis fixation has been claimed by the use of porous calcium aluminate for tissue growth into bone pores, although loss in strength with time arises. Piezoelectric barium titanate incorporating a textured surface has been investigated also to improve bone fixation. Bone healing and growth are stimulated by mechanical loading during use of the implant.

5.3.3 Applications of Ceramics in Joint Replacements

Alumina (Al_2O_3) and Zirconia (ZrO_2) ceramics are used for the femoral heads of hip replacements. They have greater stiffness, wear-resistance and hardness than metals. Calcium phosphate, such hydroxyapatite (HA) (Ca_{10}-$(PO_4)_6$-$(OH)_2$), is used to coat metallic implants in order to bond them to the bone. Tricalcium phosphate (TCP) is also used, although it degrades more quickly than HA.

Despite these uses, the brittle nature of ceramics remains a drawback. Impact, or non-uniform loading of aluminium oxide ceramics can lead to breakdown or fracture, with the release of debris. Manufacture of matching ceramic femoral heads and acetabular components to high congruency of surface can be difficult. Owing to their stiffness their deformation under load can also make congruency difficult to achieve. The area of contact between these parts of the replacement is reduced. With increased surface stress, significant wear can arise.

The articulation of ceramic with metal can lead to less wear compared to metal-on-metal (Ishida *et al.* 2009). A metal stem may have a ceramic head, although fracture of the head has been known to occur. Ceramics are stronger and less brittle in compression than in tension.

Zirconium oxide is less stiff and is stronger than aluminium oxide in tension (and also than stainless steel and the cobalt alloys) and is a tougher material. Compared to aluminium oxide, it can absorb higher loads that have tensile components without breaking. However, its hardness is less than that of aluminium oxide and it has a lower compressive strength.

In the manufacture of femoral heads for aluminium or zirconium oxides, these ceramics are first obtained in powder form. They are then sintered or hot pressed into a dense condition with small grains. They are chemically inert, are not subject to wear due to oxidation (which can lead to roughening of metal heads of prostheses), nor do they release metallic ions. A smooth surface finish can be readily obtained. For example, the coefficient of friction of aluminium or zirconium oxide on polyethylene is less than that of metal on polyethylene, and the corresponding amount of wear is between three and sixteen times lower.

Two ceramic-on-ceramic materials, alumina-toughened zirconia (ATZ-on-ATZ), and zirconia-toughened alumina (ZTA-on-ZTA), have been found to be more wear resistant than other ceramic bearing materials alumina oxide (Al-on-Al) in hip simulation conditions (Jennings *et al.* 2011). Zirconia-toughened alumina ceramic heads have also been compared with carbon-fibre reinforced polyether-ether-ketone (CFR-PEEK) cups and with metal-on-metal hip resurfacing materials (Everitt *et al.* 2011). They conclude that metal-on-metal configurations have a higher friction factor compared to ZTA and CFR-PEEK.

Park (1984), drawing attention to the similar molecular structure of HA and bone, and Park and Bronzino (2002), discuss variations in the mechanical properties of HA. Typical properties are compressive strengths from 917 to 294 MPa, tensile strengths of 196 MPa, bending strengths of 147 MPa and Vickers hardness of 3.43 MPa. The elastic modulus of the coating varies from 117 to 144 GPa; c.f. human cortical bone which is 24.6-35 GPa. The Poisson's ratio is about 0.27 for mineral or synthetic HA. That for bone is about 0.3.

Park (1984) attributes the attractive biocompatibility of HA to its direct chemical bonding with hard tissue. He suggests that osteoblast cells attach themselves to the HA that is coated to the surfaces of implants, most widely by plasma spraying (see Chapter 6), the bonding between the HA and substrate is vital. Their attachment promotes anchorage of cells, and through their adhesion to the implant, cell growth is increased. Bone tissue can then be deposited onto the material surface. The terms 'biointegration' and 'osseointegration' are applied to describe the bonding of new bone to a material such as HA. This condition appears to be enhanced by the HA surface becoming roughened by the attachment of macrophages; the osteoblasts lay down osteoids on the rough surface which then act as epitaxial nucleation sites on which apatite is formed.

For satisfactory performance of a HA coating, the interfaces between bone and HA, and HA and the implant play a key part. Park (1984) discusses cases in which the interface between Ti or Ti6Al4V alloy and the HA coating have failed, quoting a low tensile strength of approximately 6.7 MPa between the HA and the alloy.

5.4 Polymers

5.4.1 Structure

'Polymers' are molecules composed on many ('poly-') parts ('mer') joined together by chemical covalent bonds.

The term 'homopolymer' is used to describe a polymer that is made up of individual or 'monomer' parts that are all the same. Homopolymers are widely used in joint replacements, and a 'copolymer' is the name used when these individual parts are different.

Figure 5.2 shows the basic characteristics of monomers designated as 'A' parts and co-polymers (with 'A' and 'B' parts).

The molecular size of a polymer is much larger than those found in ceramics or the atoms that make up metal alloys. For example, a polymer can contain over 100 000 monomers, and have a molecular weight of millions g/mol. A simple ceramic like MgO comprises a one-to-one ratio of metallic and

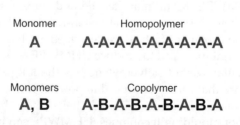

Monomer Homopolymer

A **A-A-A-A-A-A-A-A-A**

Monomers Copolymer

A, B **A-B-A-B-A-B-A-B-A**

Figure 5.2 Homopolymer and copolymer structure.

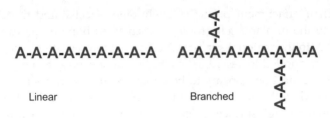

Figure 5.3 Linear and branched polymer structure.

non-metallic atoms with respectively atomic weights of 54.9 and 8. The atomic weights of chromium and cobalt employed in joint replacements made of chromium-cobalt alloys, are respectively 51.9 and 58.9 g/mol.

Figure 5.3 shows a further characteristic of polymers to remain linear or branched, dependent on how they are processed. They may also exhibit a repeating structure.

In Figure 5.4, well-known polymeric materials with such a structure are presented. In particular, polyethylene is indicated.

It has a chemical formula, of $(C_2H_4)_n$, with n denoting the degree of polymerization. This polymer is obtained from ethylene (C_2H_4), a gas with a molecular weight of 28. Their chemical structures are illustrated in Figure 5.5.

5.4.2 Ultra-high Molecular Weight Polyethylene (UHMWPE)

The joint replacement bearing material, linear homopolymer UHMWPE is composed of many (for example 200 000) ethylene repeat units, thereby containing as many as 400 000 atoms. Its average molecular weight can be as high as 6 million g/mol. This polymer can also be copolymerized with other monomers such as polypropylene to obtain required physical and mechanical properties. Typical properties are given in Table 5.10. For comparison, properties of another linear polymer of the polyethylene group HDPE (high density) are included. Its molecular weight, of typically 200 000 g/mol, is much less than that of UHMWPE. Its ultimate tensile and impact strengths are also lower, as well as being less wear- and abrasion-resistant than UHMWPE.

Although methods of manufacture are described in Chapter 6, it is useful to summarize the techniques used to produce UHMWPE for joint replacements. Ultra-high molecular weight polyethylene has the longest, least-branched molecular structure that can be produced in powder form, in large amounts. The powder is solidified into its required shape under heat and pressure by extrusion or hot moulding techniques. UHMWPE can be produced with defined mechanical properties by the use of specific methods of manufacture.

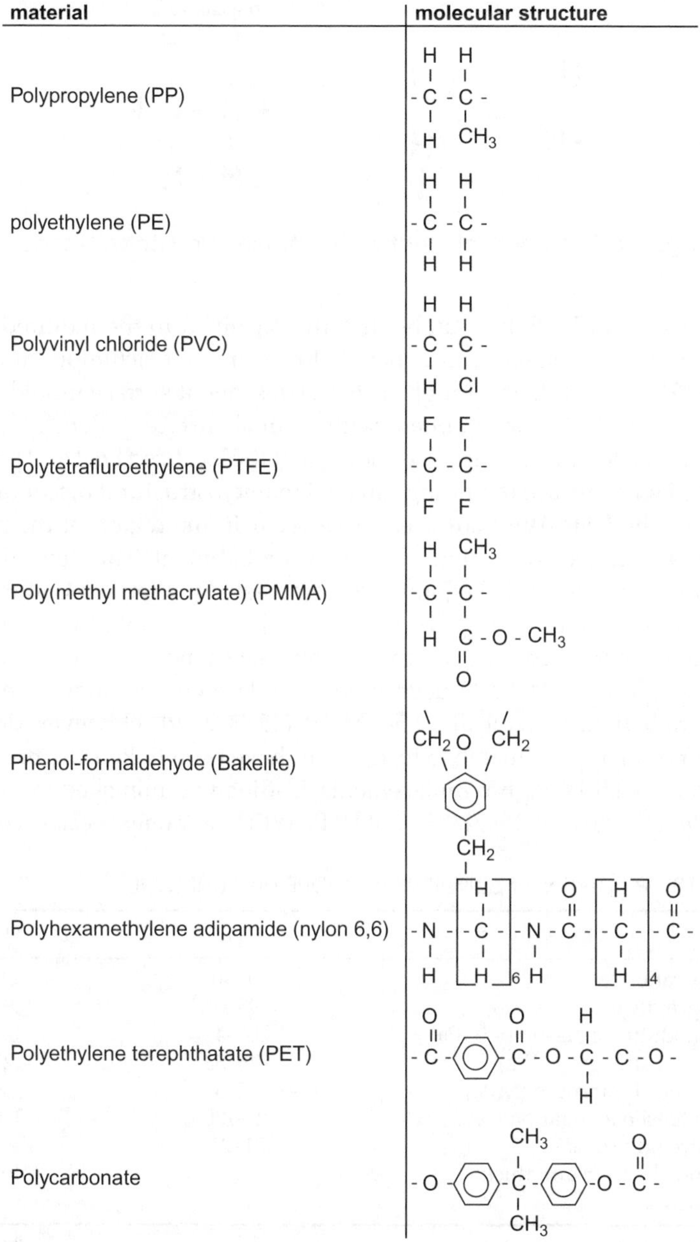

material	molecular structure
Polypropylene (PP)	
polyethylene (PE)	
Polyvinyl chloride (PVC)	
Polytetrafluroethylene (PTFE)	
Poly(methyl methacrylate) (PMMA)	
Phenol-formaldehyde (Bakelite)	
Polyhexamethylene adipamide (nylon 6,6)	
Polyethylene terephthatate (PET)	
Polycarbonate	

Figure 5.4 Repeating (mer) structures of polymers.
Source: Reproduced with permission from Callister, W.D. (1994) *Materials Science and Engineering,* 3 edn, John Wiley & Sons Inc., New York.

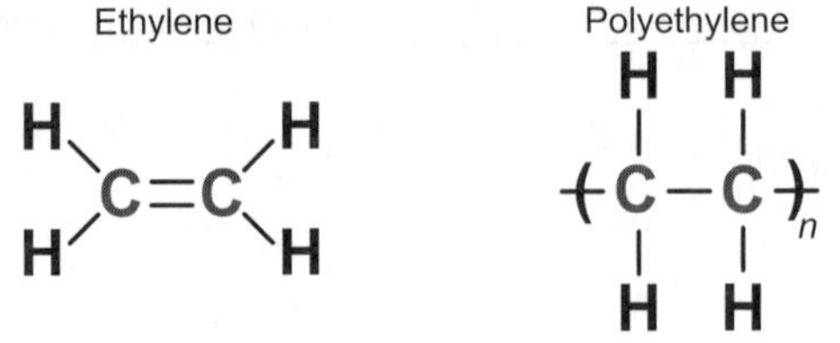

Ethylene   Polyethylene

$$\underset{H}{\overset{H}{\diagdown}}C=C\underset{H}{\overset{H}{\diagup}}$$ $$+\underset{\underset{H}{|}}{\overset{\overset{H}{|}}{C}}-\underset{\underset{H}{|}}{\overset{\overset{H}{|}}{C}}\mathbin{\big)_{n}}$$

Figure 5.5 Chemical structure of ethylene and polyethylene.

Extruded UHMWPE bar can be directly machined to the required shape of the joint replacement component. However, as machining can leave UHMWPE with a dull and rough surface, subsequent surface moulding can be used to provide the component with its final surface dimensions, and a glossy smooth texture. (It should be noted that the localized heat and pressure associated with surface moulding can lead to microstructural discontinuities just below the UHMWPE surface. These occur in the region of the highest shear stress, caused by motion of the joint replacement. This condition has been known to arise in knee joint replacements.) An alternative to machining is compression moulding, which can yield a shape close to that needed, requiring further machining to produce the final shape.

Sterilization of UHMWPE used in joint replacement is undertaken with gamma radiation, at typically 2.5–5 Mrad (25–50 kGy). However, doses of radiation from approximately 5 to 15 Mrad have been linked to progressive oxidation of UHMWPE hip replacements, leading to diminished mechanical properties (Naidu *et al.* 1997). Wear of UHMWPE components can give rise to

Table 5.10 Properties of polyethylenes HDPE and UHMWPE.

Property	HDPE	UHMWPE
Poisson's ratio	0.40	0.46
Specific gravity	0.95–0.97	0.93–0.95
Tensile modulus of elasticity (GPa)	0.4–4.0	0.8–1.6
Tensile yield strength (MPa)	26–33	21–28
Ultimate tensile strength (MPa)	22–31	39–48
Ultimate tensile elongation (%)	10–1200	350–525
Impact strength (Izod)	21–214	>1070
(J/m of notch; 3.175 mm thick specimen)		(No break)

Source: Adapted from Edidin, A.A. and Kurtz, S.M. (2000) The influence of mechanical behavior on the wear of four clinically relevant polymeric biomaterials in a hip simulator. *Journal of Arthroplasty* **15**, 321–331.

debris of micron to submicron size. Loosening of the joint replacement then arises.

Ultra-high molecular weight polyethylene is used as the acetabular cup for total hip arthroplasty, and as the tibial bearing surface for total knee replacements. It is used to replace the socket in shoulders and the radial component in total elbow replacement. The polymer can also be found in toe, finger, wrist and ankle joint implant devices (ASTM Standard F 2759–2011).

The UHMWPE currently used is less crystalline and dense than that previously used by Charnley. He first tried polytetrafluoroethylene (PTFE) before moving to high-density polyethylene. Substitution of UHMWPE by a more crystalline and denser polyethylene (commercially available as 'Hylamer'), which is stiffer, stronger and more resistant to creep, was investigated. High rates of eccentric wear have meant use of this alternative material has been largely abandoned. Hot-moulded polyethylene, reinforced with inert carbon fibres of high elastic modulus that offer greater resistance to creep and wear, and enhanced compressive strength, have been used for hip and knee components. With little significant improvement over UHMWPE the application of this material has also been discarded.

5.4.3 Polymer Cement

Polymers with no cross-linking of polymer chains, such as polyethylene and poly(methyl methacrylate) (PMMA) are extensively used in joint replacements. Pre-polymerized PMMA powder and liquid methyl methacrylate (monomer) are mixed to form acrylic bone cement. The liquid monomer possesses a double bond, and upon mixing the bond is broken. In the resulting activation reaction, free valences are able to join up with other similar molecules. The monomer is then converted into a polymer, with no side products being formed. This process is termed 'free radical addition' and occurs until long-chain molecules are obtained, in the form of a chain reaction. Tables 5.11 and 5.12 show typical constituents of bone cement. The rate of curing of the solid cement can be increased by the addition of N,N-Dimethyl-p-toluidine (DMPT). This high-production volume chemical has drawn controversial comments, as it has been suspected of possessing carcinogenic properties, as discussed by Lewis (2007). Figure 5.6 illustrates the structure of the bone cement after curing. The polymer powder particles are wetted by the monopolymer liquid, which is polymerized, becoming solid. The liquid also connects them together after polymerization.

Typical times of mixing and setting are respectively 5 minutes, and 5 to 15 minutes for the cement powder-liquid mixture. A minimum compressive

Table 5.11   Constituents of bone cement.

Liquid component (20 ml)	
Methyl methacrylate (monomer)	97.4 vol%
N,N-Dimethyl-p-toluidine	2.6 vol%
Hydroquinone	75 ± 15 ppm
Solid powder component (40 g)	
Poly(methyl methacrylate)	15.0 wt%
Methyl methacrylate-styrene copolymer	75.0 wt%
Barium sulphate (BaSO$_4$), USP (meets medical grade)	10.0 wt%
Dibenzoyl peroxide	Not available

Source: Adapted from Howmedica, Inc. (1977) *Safety Data Sheet-Surgical Simplex(r) Liquid, Stryker,* Limerick, Ireland.

strength of 70 MPa is expected. Typical molecular weights of the constituents of the monomer, powder, and the cement after curing are given in Table 5.13.

The mechanical properties of bone cement are affected by the formation of pores (sometimes up to a diameter of 1 mm) during curing. These properties can also be influenced by other factors including the ratio of liquid to powder, particle size, temperature and humidity.

The cement is used to ensure that implants are fixed securely to neighbouring bone. The cement affects the distribution of load between bone and

Table 5.12   Further details of constituents of bone cement.

Powder	
Beads of pre-polymerized PMMA poly(methyl methacrylate)-based polymer	83–99 wt %
Radiopacifier	
BaSO$_4$ or ZrO$_2$ particles	9–15 wt %
Initiator of polymerization	
Benzoyl peroxide (BPO)	0.75–2.6 wt %
Liquid monomer	
Methyl methacrylate monomer	97–99 wt %
Accelerator of polymerization reaction	
N,N-Dimethyl-p-toluidine (DMPT)	0.4–2.8 wt %
Inhibitor of that reaction	
hydroquinone	15–75 ppm

Source: After Lewis, G. (2007) Alternative acrylic bone cement formulations for cemented arthroplasties: present status, key issues, and future prospects. *Journal of Biomedical Materials Research Part B: Applied Biomaterials* **84B**, 301–319.

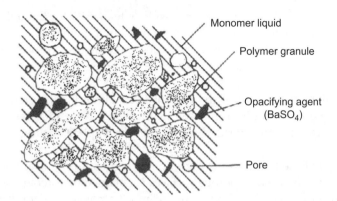

Monomer liquid

Polymer granule

Opacifying agent
($BaSO_4$)

Pore

Figure 5.6 Illustration of structure of bone cement after curing.
Source: Reproduced from Park, J.B. and Lakes, R.S. (1992) *Biomaterials: An Introduction*, 2nd edn, Plenum Press, New York.

implant. As the cement hardens, its penetration into the bone provides a load-bearing interlock, which stabilizes the implant.

Poly(methyl methacrylate) (PMMA) cement is used to fix prostheses securely to adjoining bone in total joint arthroplasty. It also effects uniform distribution of load away from the component surface to that of the bone, in order to reduce stress on the supporting bone. Judicious mixing and handling of the PMMA is a key to its effectiveness in supplying a durable interface between the surface of implant and the bone, to avoid loosening.

The cement is cold curing. (Note that the reaction is exothermic, so the cement temperature does become briefly raised.) Heat and pressure are not required, as the substance contains a polymerization initiator in its powder and an activator in its liquid, which are duly mixed. Prior to mixing, the cement is composed of two main elements: PMMA powder of particles 10 to 150 μm in diameter, about 10% of radiopaque barium sulphate or zirconium

Table 5.13 Molecular weight (M.W.) of monomer, powder and cured cement.

Type of M.W. (g/mol)	Monomer	Powder	Cured
M_n (number average)	100	44 000	51 000
M_w (weight average)	100	198 000	242 000

Source: After Haas, S.S., Brauer, G.M. and Dickson, G. (1975) A characterization of polymethyl-methacrylate bone cement. *Journal of Bone Joint Surgery* **57A**, 280–291.

dioxide and about 1% of benzoyl peroxide, which initiates the polymerization. The second item is a liquid mixture of methyl methacrylate monomer,
3% DMPT, and an activator which facilitates the cold-curing process. (The liquid also has a trace of a retardant to reduce monomer polymerization during
storage.)

Preparation and setting of the cement takes about eight to ten minutes.
Initial mixing to a dough form accounts for two to three minutes, followed
by five to eight minutes needed to position the implant securely, and then
setting.

Poly(methyl methacrylate) is a grout, not a glue. Mechanical bonding to
implants with textured or porous surfaces is obtained by forcing the PMMA
cement into the spaces of the texture or into the pores. The bond provides
respectively greater and lower resistance to shear and tensile force at the
surface.

When cancellous bone is to be secured to the implant, the PMMA in its
low viscosity state has to be forced into the interstices of the bone. In these
circumstances a secure mechanical bond prevents motion at the interface
of the bone and cement. Avoiding any motion at the interface is crucial in
preventing poor load transfer and production of debris arising from wear,
and hence loosening.

Poly(methyl methacrylate) is about three times stronger in compression
than in tension, and can therefore deal effectively with compressive loading.
As it is brittle when polymerized, it will produce stress risers that could lead
to fracture if used with implants with sharp edges.

Metallic implants are sometimes precoated with a thin surface layer of
PMMA during manufacture in order to enhance the fixation process, through
chemical bonding of the surface to the cement. The properties of PMMA
cement are discussed in detail by Lautenschlager *et al.* (1984).

In a more recent review, Lewis (2007) observes that all commercially available plain acrylic bone cements used in arthroplasties rely on poly(methyl
methacrylate) (PMMA), mostly also with a similar composition. More than
30 types have been approved by the regulatory bodies (see Chapter 8). Their
reliability is well established: 94–96% of cements last with total hip joint
replacements that have been fitted for 10 years. On the other hand, Lewis
draws attention to the deficiencies of such cements, such as high exothermic
temperature, lack of bioactivity, and volumetric shrinking on curing. Reser-
vations are expressed about some constituents of these cements, especially
toxicity effects of the activator *N,N*-Dimethyl-*p*-toluidine (DMPT), discussed
above, and wear related to radiopacifiers $BaSO_4$ or ZrO_2 particles. Alternatives that avoid or diminish these deleterious conditions have been sought
for more than 30 years. Lewis divides alternative cement formulations into

16 groups. Each is designed to tackle one or more of the deficiencies of the established cements. He suggests that most continuing development lies in an alternative radiopacifier, bioactive, and hydrophilic partially degradable bioactive classes (HPDBC).

The mechanical and biological deficiencies of PMMA-based bone cement have been investigated by Khandaker *et al.* (2011). In their experiments, they studied additives and alternative monomers that might improve the PMMA cement characteristics. They conclude that the flexural strength and fracture toughness of the cobalt HV PMMA bone cement can be increased significantly when the monomer glycidyl methacrylate is used in addition to, or to replace, the methyl methacrylate (MMA) monomer.

5.5 Joint Replacement Materials in Service

The materials used in joint replacement can be affected by wear and friction, fatigue and creep, and corrosion.

5.5.1 Wear and Friction

The loss of material from a surface constitutes 'wear'. The main types are adhesive, abrasive and fatigue wear. Wear may also arise from other effects such as erosion, fretting, impact and cavitation.

In joint replacement, the bond associated with an adhesive can be stronger than the underlying material. On application of a tangential force, fracture may occur in the softer material, giving rise to wear fragments, which upon detachment during subsequent rubbing become loose wear particles. Closely associated with wear is 'friction' – that is the resistance to relative sliding between two bodies in contact under loading. The energy dissipated in over-coming friction is converted into heat which in turn increases the temperature at the interface between two bodies.

Fatigue wear is caused by cyclic loading, which induces crack propagation adjacent to the surface. The meeting of two or more cracks usually gives rise to removal of fragments of material or wear particles, known as spalling or pitting. Wear due to corrosion is associated with oxidation of the surface of an implant, followed by consequent removal of the corrosion debris, generating wear by abrasion at the interface surface. The deleterious effect of corrosion on implants is discussed more fully in section 5.5.3.

Fisher *et al.* (2009) have investigated the effects of cross-shear and bearing surface area to reduce polyethylene wear. O'Connor *et al.* (2009) quote Wroblewski (1985) as calculating a linear wear penetration rate of 0.2 mm per year for Charnley hip replacements in the elderly, based on one-million cycles of movement for one year. An alternative material to UHMWPE is carbon-fibre reinforced poly-ether-etherketone (CFR-PEEK), which has been found to have reduced volumetric wear rates compared with established metal-on-UHMWPE unicondylar knee joints (Scholes and Unsworth 2009). The same CFR-PEEK material exhibited lower wear in hip prostheses (Scholes *et al.* 2008). Minimal adverse reaction by tissue to this material has also been indicated by Howling *et al.* (2003). For metal-on-metal acetabular cups, methods of calculating wear location paths and depth by use of 'out-of-roundness' machines have discussed by Morris *et al.* (2011).

Cobalt-chromium-molybdenum alloy (ASTMF75), based on the cutting tool material Stellite 21, is also used in the manufacture of hip bearings. It has been reported that cobalt-chromium nanosized particles are released from the surface being generated by metal-on-metal articulations.

Concerns have been expressed over inflammatory tissue response and risks of cytotoxicity, with wear from these metal-on-metal hip bearings. (Cytotoxicity refers to poisonous effects on cells.) De Wet *et al.* (2011) have suggested that another Co-Cr-Mo alloy, Stellite 720, shows lower wear rates compared with presently used alloys of the Stellite 21 type.

Closely associated with wear is friction, that is, the resistance to relative sliding between two bodies in contact under loading. Examples of friction have been discussed above, for instance in section 5.2.2. and later in Chapter 8.

5.5.2 Fatigue and Creep

Repeated or cyclic stressing on a material can cause it eventually to fail, even at a stress well below its fracture level. This condition is known as 'fatigue failure'. Materials can be tested for fatigue by application of alternating stress. Steels exhibit a precise fatigue limit, unlike nonferrous materials where there is usually no definite fatigue limit. Fatigue strength is affected by surface condition, design and the implant environment. In joint replacement materials, the condition can be associated with corrosion as noted below.

In contrast to fatigue conditions, when a constant load is applied to a material it can undergo continued slow straining, known as 'creep' the extent of which is affected by both time and temperature. Polymers can be affected by creep, although if the stress on them is kept low the condition can be avoided.

Connelly *et al.* (2005) performed cyclic loading of plain and carbon-fibre reinforced UHMWPE used in total knee replacements in order to establish fatigue crack propagation patterns. They found that fatigue crack resistance in the reinforced UHMWPE was worse than that of the plain UHMWPE.

They attributed this result to insufficient bonding between the carbon fibres and the UHMWPE.

Lennon *et al.* (2007) used patient-specific finite element analysis to simulate creep and fatigue damage in order to anticipate failure of hip joint prostheses. Their analysis concluded that computational simulation of loading over a ten-year period during normal gait activities could be used in preoperative assessment.

5.5.3 Corrosion

The term 'corrosion' is used to describe the deterioration of metals, usually by chemical or electrochemical attack. Bodily fluids are known to corrode the materials from which joint replacements are manufactured. The fatigue life and strength of the implant material may then be reduced, and mechanical failure may ensue.

Useful indications of corrosion behaviour can be derived from the tables of standard electrode potentials derived from electrochemical free energy equations (McGeough 1974). However this electromotive series does not account for the passive oxide films that form on some of the alloys commonly used in joint replacement. Neither does it account for the condition of these alloys when they are in contact not with a standard 'normal' electrolyte, but with a more bodily fluid type of solution. The well-known galvanic series of metal behaviour in sea water is a preferable representation. As noted in Table 5.14 materials towards the cathodic end are more protected from corrosion than those in the upper 'anodic' region. The position of the orthopaedic materials of passive stainless steel and titanium towards the cathodic end confirms

Table 5.14 Galvanic series of implant metals.

Anodic/end:	Magnesium alloys
	Zinc
	Galvanized steel
	Aluminium
	Mid steel
	Austenitic stainless steel (active)
	Nickel
	Copper
	70/30 Cupronickel
	Austenitic stainless steel (passive)
	Titanium
Cathodic end:	Gold

that the oxide layer formed on these metals renders them suitable for joint replacements. Nonetheless these oxide films have to be non-porous, adhere to, and extend over, the entire metal surface, and be able to withstand the effects of mechanical stress and abrasion.

However, corrosion of implants can occur, the most common of which are:

- **Corrosion fatigue** – when the material undergoes alternating and fluctuating load cycles. Bodily fluids can cause the formation and development of cracks, thereby reducing the fatigue life of a material. Charnley (1970) attributed a single failure in the stainless steel component of a hip replacement to corrosion fatigue, stemming from corrosion pitting.
- **Corrosion pitting** – this effect can arise from localized breakdown of the protective film on the alloy surface. Pitting can occur at a discontinuity in the oxide layer and is more prevalent with highly resistant films. The latter limits the extent of corrosion over wider regions so that the effect is confined to a series of localized areas. The mechanisms underlying corrosion pitting and fatigue are illustrated in Figure 5.7.
- **Stress-corrosion** – cracks initiated due to the combined effects of stress and a corrosive environment can subsequently propagate through an alloy given rise to stress corrosion.
- **Crevice corrosion** – cracks can give rise to long, narrow crevices with low concentrations of oxygen. Reports on the fracture of stainless steel stems have been attributed to a breakdown of cement in the proximal part of the implant stem. This failure appears to have been due to abrasion of the film on the stainless steel stem caused by motion between stem and cement.

The corrosion of metal orthopaedic implants has been discussed in detail by Jacobs *et al.* (1998). For example, they report corrosion signs in the head

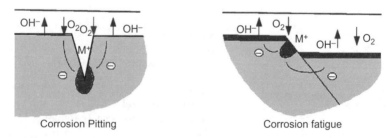

Corrosion Pitting   Corrosion fatigue

Figure 5.7   Examples of corrosion in metals.

to neck taper connection between a titanium alloy femoral stem and a cobalt alloy femoral head. Corrosion was also evident when a combination of cobalt-alloy stems and cobalt-alloy heads was used. They also describe intergranular corrosion in cobalt alloys and discuss evidence for interface crevice corrosion in stainless steel internal fixation devices. They draw attention to local tissue effects including osseointegration and fibrous encapsulation, as well as tissue reaction to the presence of foreign bodies such as polyethylene and bone cement debris.

Although polymers and ceramics would not be expected to be affected by corrosion, they can be susceptible to other forms of chemical deterioration. Wang *et al.* (1996) have investigated the mechanisms of wear in UHMWPE joint replacement prostheses. They draw attention to changes in the molecular chain structure at the articulating surface of UHMWPE due to strain, causing a fibre-like surface texture. They discuss the wear mechanism involving failure between the fibres rather than that caused by the inherent strength of the fibres themselves. Fluctuating stresses have been known to cause fatigue, leading to high wear rates of UHMWPE.

In hip and knee bearings, ceramic surfaces offer major reductions in rates of wear with attractive biocompatibility. Rahaman *et al.* (2007) have reviewed the place of ceramics in the replacement of these joints, although they do draw attention to the major advances gained with the use of ceramics, as well as reporting some cases of failure *in vivo*. They suggest that if ceramics do fail *in vivo*, this is due to either flaws in the fabrication process or corrosion degradation, leading to slow growth of cracks. They also summarize the qualities of the ceramic materials that are either currently in use or are under development for bearings in total hip and knee replacements.

5.6 Nanomaterials

There is no international standard that defines nanoengineering or its technologies. However, this branch of engineering is concerned with materials and their related processes and systems that involve dimensions of about 1 to 100 nm. The design of nanoengineering processes is concerned with the influence or control of the mechanical, electrical and chemical characteristics at a molecular scale. From these bases larger structures may be produced (Corbett *et al.* 2000).

Nanoscale structures can be studied by scanning electron microscopes (SEM). Their surfaces can also be inspected using a scanning probe microscope, which features a probe of tip radius of about 10 nm. Atomic force microscopes (AFM) perform the same task, and can also be employed to manipulate or move even single atoms or larger nanoscale objects. Nanoscale shapes are produced by means of processes such as photo- and X-ray lithography and focused ion-beam machining. Materials of shape or grain size less than 100 nm are receiving increased attention, owing to their attractive mechanical, electrical and other properties. Highly wear-resistant coatings for use in joint replacements can be manufactured from such nanomaterials.

The production of nanotube-like structures from boron nitride, molybdenum and carbon, with concentric layers of diameters from about 0.4 nm and lengths of 1 μm have received particular study, especially carbon nanotubes (CNT). Chemical vapour deposition (CVD), discussed in Chapter 6, is one method used in the synthesis of CNT.

Exceptional mechanical properties have been reported by Mamalis *et al.* (2004) for CNT, with Young's moduli of elasticity estimated at about 1 to 5 TPa, tensile strengths of 11 to 63 GPa (cf. high strength steels, breaking at 2 GPa), and densities of 1330 to 1400 kg/m^3 (cf. aluminium at 2700 kg/m^3).

Nanotubes are claimed to offer almost wear-free surfaces, with immediate relevance to joint replacements. In addition nano-mechanical or nano-electromechanical systems (NEMS) in which low-friction/low-wear nano-sized bearings and nano springs are feasible become an attractive advancement to general engineering.

Closely associated with these advances in nanotechnological materials are nanocomposites. These are solid materials composed of more than one phase, in which the dimensions of one of these phases are less than 100 nm. The mechanical, electrical and other properties of nanocomposites have significant differences compared to other established composite materials. These contrasting characteristics arise from the very high surface to volume, or aspect, ratios of the 'nanophase' reinforcement materials. The latter materials can be composed of particles, or fibres, such as CNT, and are distributed in the matrix material during processing. Although even the mass fraction, or percentage by weight, of the nano-particles is only about 0.5 to 5%, the large surface area of reinforcement that they cover gives rise to major changes in macroscopic properties of the nanocomposite material. In consequence, mechanical properties including wear resistance, stiffness and strength can be improved, as well as electrical, thermal and optical characteristics.

For joint replacements, ceramic- and polymer-matrix nanocomposites are receiving interest. With the former, a ceramic occupies the bulk of its volume with usually a metal providing the secondary constituent. Ceramic-matrix nanocomposites have enhanced resistance to corrosion, and other improved qualities of protection. In polymer-matrix nanocomposites, the polymer matrix has added to it nanoparticles such as ceramics or CNT which have substantially different properties; for instance they can enhance stiffness or features such as biodegradability.

The suitability of CNT and polymer nanocomposites as scaffold materials for bone cell growth has been reported by Shi *et al.* (2007) and Sithamaran *et al.* (2008).

A novel nanocomposite material composed of hydroxyapatite (HA) modified with carbon-nanotube-reinforced poly(methyl methacrylate) (PMMA) has been prepared by Singh *et al.* (2008). They draw attention to the attractions of the combination of mechanical, thermal and electronic properties of multi-walled carbon nanotubes. Their combination of PMMA and HA is proposed as offering enhanced qualities as bone cement. Only an addition of 0.1% concentration in weight of the CNT material to the PMMA modified HA is claimed to give superior mechanical properties that could overcome difficulties encountered with the brittleness and inadequate strength of sintered HA and with the mechanical interlocking fixation strength of PMMA cement to bone.

The new material was fatigue tested on artificial bone over one-million cycles without evidence of crack propagation or delamination, which were found in similar tests on nanocomposites containing only PMMA.

Curtis *et al.* (2006) have undertaken extensive studies on the effects of nanoscale topography on stem cells with implications for bone-tissue engineering.

A second-generation highly cross-linked polyethylene with enhanced strength and wear resistance, developed by Harris, has been used by Freiberg (2011) to line the joint socket in total hip replacement. The material is particularly suitable when the replacement hip has to withstand the higher stress conditions encountered with younger, more active and also heavier people receiving a replacement.

Torrecillas *et al.* (2009) have reviewed recent achievements in nanomaterials and related processes for joint replacement. They discuss methods for improving the osseointegration of implant to bone, and to the sensitivity of osteoblasts to the roughness of the material surface on which they are deposited. Attention is drawn to drawbacks associated with loosening and failure of prostheses due to inflammatory reaction of the body to the implantation of materials that are artificial to it. These adverse reactions may be reduced by improved protective coatings. Further materials used to provide biocompatibility with bone metal implants include tricalcium phosphate (TCP). Coatings are about 50 microns in thickness and are hot plasma sprayed onto the implant. Although HA coatings are more common, and little deterioration arises even after lengthy periods, the rate of absorption of HA is slow. On the other hand, TCP absorbs more rapidly and is used to promote early bone growth into porous surfaces.

References

Ajayan, P.M., Schadler, L.S. and Braun, P.V. (2003) *Nanocomposite Science and Technology*, John Wiley & Sons, Ltd, Chichester.

American Society for Testing and Materials Standards (1980) *Annual Book of ASTM Standards*, American Society for Testing and Materials Standards. Philadelphia, PA.

ASTM Standard F2759 (2011) *Standard Guide for Assessment of the Ultra High Molecular Weight Polyethylene (UHMWPE) Used in Orthopedic and Spinal Devices*, ASTM International, West Conshohocken, PA.

Black, J. (1988) *Orthopaedic Biomaterials in Research and Practice*, Churchill Livingstone, New York.

Callister, W.D. (1994) *Materials Science and Engineering*, 3rd edn, John Wiley & Sons, Inc., New York, Ch. 15.

Charnley, J. (ed.) (1970) Total hip replacement. *Clinical Orthopaedics and Related Research* **72**, 1–241.

Choi, K., Kuhn, J.L., Ciarelli, M.J. and Goldstein, S.A. (1990) The elastic moduli of human subchondral, trabecular, and cortical bone tissue and the size-dependency of cortical bone modulus. *Journal of Biomechanics* **23**, (11), 1103–1113.

Connelly, G.M., Rimnac, C.M., Wright, T.M. *et al.* (2005) Fatigue crack propagation behaviour of ultra-high molecular weight polyethylene. *Journal of Orthopaedic Research* **2**, 119–125.

Corbett, J., McKeown, P.A., Peggs, G.N. and Whatmore, R. (2000) Nanotechnology: international developments and emerging products. *Annals of the CIRP* **49** (2), 523–545.

Curtis, A.S.G., Dalby, M. and Gadegaard, N. (2006) Cell signalling arising from nanotopography: implication for nanomedical devices. *Journal of Nanomedicine* **1** (1), 67–72.

De Wet, D.J., Yao, M., Collier, R. *et al.* (2011) New Co-Cr alloys with improved wear properties for metal-on-metal hip bearings. Institution of Mechanical Engineers (IMechE), event proceedings, Engineers and Surgeons: Joined at the Hip III, p. 57.

Edidin, A.A. and Kurtz, S.M. (2000) The influence of mechanical behavior on the wear of four clinically relevant polymeric biomaterials in a hip simulator. *Journal of Arthroplasty* **15**, 321–331.

Everitt, H., Elliott, M., Bigsby, R. *et al.* (2011) Comparison between friction and lubrication behaviour of large diameter ZTA ceramic vs CFR-PEEK and MOM hip resurfacing implants. Institution of Mechanical Engineers (IMechE), event proceedings, Engineers and Surgeons: Joined at the Hip III, p. 79.

Fisher, J., Jennings, L.M., Galvin, A. *et al.* (2009) Wear of polyethylene in total knee replacements. Institution of Mechanical Engineers (IMechE), event proceedings, Knee Arthroplasty 2009: From Early Intervention to Revision, pp. 85–87.

Freiberg, A.A., Kwon, Y-M., Malchau. H. *et al.* (2011) Management of polyethylene wear associated with a well-fixed modular cementless shell during revision total hip arthroplasty. *Journal of Arthroplasty* **26** (4), 576–581.

Haas, S.S., Brauer, G.M. and Dickson, G. (1975) A characterization of polymethylmethacrylate bone cement. *Journal of Bone Joint Surgery* **57A**, 280–291.

Hille, G.H. (1966) Titanium for surgical implants. *Journal of Materials* **1**, 373–383.

Howling, G.I., Sakoda, H., Antonarulrajah, A. *et al.* (2003) Biological response to wear debris generated in carbon based composites as potential bearing surfaces for artificial hip joints. *Journal of Biomedical Materials Research Part B: Applied Biomaterials* **67** (2), 758–764.

Howmedica, Inc. (1977) *Safety Data Sheet- Surgical Simplex(r) Liquid*, Stryker, Limerick, Ireland.

Ishida, T., Clarke, I.C., Donaldson, T.K. *et al.* (2009) Comparing ceramic-metal to metal-metal total hip replacements – a simulator study of metal wear and ion release in 32- and 38-mm bearings. *Journal of Biomedical Materials Research Part B: Applied Biomaterials* **91** (2), 887–896.

Jacobs, J.J., Gilbert, J.L. and Urban, R.M. (1998) Current concepts review – corrosion of metal orthopaedic implants. *Journal of Bone and Joint Surgery* **80**, 268–282.

Jennings, L.M., Al-Hajjar, M., Begand, S. *et al.* (2011) Wear of novel ceramic-on-ceramic bearings under adverse and clinically relevant hip simulator conditions. Institution of Mechanical Engineers (IMechE), event proceedings, Engineers and Surgeons: Joined at the Hip III, pp. 81–83.

Khandaker, M., Liu, P., Li, Y. and Vaughan, M.B. (2011) Bioactive additives and functional monomers effect on PMMA cement: mechanical and biocompatabilty properties. American Society of Mechanical Engineers (ASME), event proceedings, 2011 International Mechanical Engineering Congress and Exposition, Denver, CO, pp. 1–7.

Kurtz, S.M. (2011) *PEEK Biomaterials Handbook*, William Andrew, Norwich.

Lautenschlager, E.P., Stupp, S.I. and Keller, J.C. (1984) Structure and properties of acylic bone cement, in *Functional Behaviour of Orthopaedic Biomaterials*, Vol. 11, (eds P. Ducheyne and G. Hastings), CRC Press, Boca Raton, FL, U.S.A.

Lennon, A.B., Britton, J.R., MacNiocaill, R.F. *et al.* (2007) Predicting revision risk for aseptic loosening of femoral components in total hip arthroplasty in individual patients – a finite element study. *Journal of Orthopaedic Research* **25** (6), 779–787.

Lewis, G. (2007) Alternative acrylic bone cement formulations for cemented arthroplasties: present status, key issues, and future prospects. *Journal of Biomedical Materials Research Part B: Applied Biomaterials* **84B**, 301–319.

Li, Z., Gu, X., Lou, S. and Zheng, Y. (2008) The development of binary Mg-Ca alloys for use as biodegradable materials within bone. *Biomaterials* **29**, 1329–1344.

Mamalis, A.G., Vogtländer, L.O.G. and Markolpoulos, A. (2004) Nanotechnology and nanostructured materials: trends in carbon nanotubes. *Precision Engineering* **28**, 16–30.

McGeough, J.A. (1974) *Principles of Electrochemical Machining*, Chapman & Hall, London.

Morris, B.O., Simpson, D.J., Collins, S.N. and Shelton, J.C. (2011) The effect of radial clearance on the location of the wear scar in metal-on-metal acetabular cups. Institution of Mechanical Engineers (IMechE), event proceedings, Engineers and Surgeons: Joined at the Hip III, p. 55.

Naidu, S.H., Bixler, B.L. and Moulton, M.J.R. (1997) Radiation-induced physical changes in UHMWPE implant components. *Journal of Orthopedics* **20** (2), 137–142.

O'Connor, J., Murray, D., Collins, M. and Blunn, G. (2009) Wear testing of the Lateral Oxford Unicompartmental Replacement. Institution of Mechanical Engineers (IMechE), event proceedings, Knee Arthroplasty 2009: From Early Intervention to Revision, pp. 97–100.

Park, J.B. (1984) *Biomaterials Science and Engineering*, Plenum Press, New York.

Park, J.B. and Bronzino, J.D. (eds) (2002) *Biomaterials: Principles and Applications*, CRC Press, Boca Raton, FL.

Park, J.B. and Lakes, R.S. (1992) *Biomaterials: An Introduction*, 2nd end, Plenum Press, New York.

Rahaman, M.N., Yao, A., Bal, B.S. *et al.* (2007) Ceramics for prosthetic hip and knee joint replacement. *Journal of American Ceramic Society* **90** (7), 1965–1988.

Scholes, S.C., Inman, I.A., Unsworth, A. and Jones, E. (2008) Tribological assessment of a flexible CFR-PEEK acetabular cup articulating against an alumina femoral head. *Proceedings of the Institution of Mechanical Engineers Part H – Journal of Engineering in Medicine* **222**(H3), 273–283.

Scholes, S.C. and Unsworth, A. (2009) Pitch-based CFR-PEEK-OPTIMA assessed as a bearing material in a mobile bearing unicondylar knee joint. Institution of Mechanical Engineers (IMechE), event proceedings, Knee Arthroplasty 2009: From Early Intervention to Revision, pp. 105–109.

Shi, X., Sitharaman, B., Pham, Q.P. *et al.* (2007) Fabrication of porous ultra-short single walled carbon nanotube composite scaffolds for bone tissue engineering. *Journal of Biomaterials* **28**, 4078–4090.

Singh, M.K., Shokuhfar, T., de Almeida Gracio, J.J. *et al.* (2008) Hydroxyapatite modified with CNT-reinforced polymethyl methacrylate: a nanocomposite material for biomedical applications. *Advanced Functional Materials* **18** (5), 694–700.

Sitharaman, B., Shi, X., Walboomers, X.F. *et al.* (2008) In vivo biocompatibility of ultra-short single walled carbon nanotube/biodegradable polymer nanocomposites for bone tissue engineering. *Bone* **43** (2), 362–370.

Staiger, M.P., Pietak, A.M., Huadmai, J. and Dias, G. (2006) Magnesium and its alloys as orthopaedic biomaterials: a Review. *Biomaterials* **27**, 1728–1734.

Torrecillas, R., Moya, J.S., Diaz, L.A. and Lopez-Esteban, S. (2009) Nanotechnology in joint replacement. *Nanomedicine and Nanobiotechnology* **1**, 540–552.

Wang, A., Stark, C. and Dumbleton, J.H. (1996) Mechanistic and morphological origins of ultra-high molecular weight polyethylene wear debris in total joint replacement prostheses. *Proceedings of the Institution of Mechanical Engineers Part H: Journal of Engineering in Medicine* **210** (3), 141–155.

Wroblewski, B.M. (1985) Direction and rate of socket wear in Charnley low-friction arthroplasty. *Journal of Bone and Joint Surgery* **67** (5), 757–761.

Zhang, S., Sun, D., Fu, Y. and Du, H. (2003) Recent advances of super hard nanocomposite coatings. *Surface Coating Technology* **167**, 113–119.

6

Methods of Manufacture of Joint Replacements

6.1 Introduction

The materials discussed in the previous chapter have to be manufactured to the shape, tolerance and surface finish required for their eventual insertion into the body as joint replacements.

Their production presents a formidable task. Traditional methods for their manufacture remain the mainstay but recently developed processes and computer-aided techniques are continually being added. This is a rapidly changing field. In this and the following chapter, procedures that are adopted to manufacture the replacements are therefore discussed.

In some cases, the same manufacturing method can be used for different materials, whether metal or ceramic or polymer. In others, choice is more restricted. In the course of their production, the surface finish, tolerance and accuracy required have to be achieved. As wear of the implant and the need for awareness of the causes and prevention of friction at the joints remain of concern prior to, and after, joint replacement, these characteristics are also described.

The Engineering of Human Joint Replacements, First Edition. J.A. McGeough.
© 2013 John Wiley & Sons, Ltd. Published 2013 by John Wiley & Sons, Ltd.

6.2 Surface Finish

Prostheses need to have a well-defined surface finish of texture before insertion. An explanation of these characteristics may therefore be useful. Finish may be described in terms of:

- waviness, a recurrent deviation from flatness of the surface, measured in terms of the width and height of the wave, and
- roughness, closely spaced irregular deviations, the scale of which is below that of its waviness. This characteristic is measured by the height, width and length (distance) on the surface of the implant.

Two other characteristics of surfaces are

- 'lay', which is the usually discernible direction of a predominant surface pattern; and
- flaws; these are random irregularities, for instance scratches, tears or cracks on the surface.

Figure 6.1 illustrates these features of a surface.

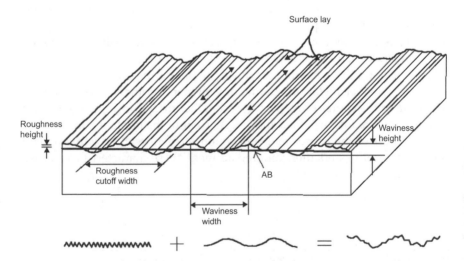

Figure 6.1   Surface roughness.

Roughness may be measured in two ways. Firstly by an arithmetic average roughness R_a (μm):

$$R_a = \frac{a + b + c + d \dots}{n}$$

where a, b, c, and d are profile heights, and n is the number of data points used. The roughness R_a is also known as centreline average (CLA). Secondly, by a root mean square average, given by R_q (μm):

$$R_q = \frac{\sqrt{a^2 + b^2 + c^2 + \dots}}{n}$$

In Figure 6.1, the line AB is defined such that the sum of the areas above and below it is equal.

Surface roughness can be measured by instruments ranging from scanning electron and optical microscopes, stereoscopic photography to profilometers.

Roughness measurements, although useful, may not represent a fully comprehensive view of a surface. Even with identical R_a and R_q values, the topography of surfaces may be different: a condition often caused by wear.

Typical surface roughnesses obtained by manufacturing processes used in the production of human joint replacement are given in Table 6.1.

Table 6.1 Typical surface roughness obtained by various processes used in the manufacture of joint replacements.

	Surface roughness (μm)
Milling	25–0.2
Grinding	6.3–0.025
Turning	25–0.05
Planing	12.5–0.4
ECM	12.5–0.05
ECG	6.8–0.2
EDM	12.5–0.8
Electropolishing	1.6–0.012
Polishing	0.8–0.012
Lapping	0.8–0.012
Honing	1.6–0.025

6.3 Tolerance

Often associated with the surface roughness that can be achieved, is the 'tolerance', which is the permissible variation in the dimension of the part that is produced. Chapter 8 describes the subject of tolerances in implants and the associated regulations.

6.4 Wear and Friction

The effects on materials of wear have already been discussed in Chapter 5. In the manufacture of implants, the consequences of potential wear have to be taken into account at the outset. The model for wear established by Archard (1953) remains a starting-point in many designs of joint replacements.

6.5 Machining

When material has to be removed from the workpiece, the technique of machining can be undertaken in various ways, as follows.

6.5.1 Milling

Figures 6.2 and 6.3 show the main features of milling, in which a multitoothed cutting tool rotates around a fixed axis.

Milling is performed with a tool, the main features of which are its (i) hardness, especially at high temperatures arising through cutting of other alloys,

Figure 6.2   Milling with multi-toothed cutting tool.

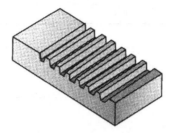

Figure 6.3 Example of feature produced by milling.

(ii) toughness, so that it does not chip or fracture, (iii) wear resistance, in order to achieve a satisfactory useful length of life for the tool, and (iv) chemical inertness, needed to avoid reactions between tool and workpiece materials.

The materials used for the cutting tool are usually:

- Carbon or medium alloy steel. These tools are inexpensive, are readily shaped and sharpened, and are generally used for low-speed cutting, as otherwise their hardness and wear resistance decrease with a rise in operating temperature.
- High-speed steels. They are also inexpensive, are wear resistant, and can be hardened to useful depths. They are tough and resistant to fracture. Their main alloying elements are molybdenum (about 10%), with chromium, vanadium, tungsten and cobalt, or tungsten (12–18%) with chromium, vanadium and cobalt. The former have higher abrasion resistance, and account for 95% of all high-speed cutting tools.
- Cast cobalt alloys, usually termed 'stellites'. These contain cobalt (38 to 53%), chromium (30 to 33%), and tungsten (10 to 20%). They have high hardness (58–64 Rockwell C scale (HRC)), can be maintained at high temperatures, and are resistant to wear.
- Carbides. For example, boron carbon nitride (BCN) and cubic boron nitride (CBN), which are used to machine chrome-cobalt alloys used in implants.

Other materials used for milling cutting tools include ceramics, silicon nitride and diamond. Some tools have coatings such as titanium nitride (TiN) and ceramic aluminium oxide (Al_2O_3).

Typical cutting speeds for milling of the materials used in joint replacement are given in Table 6.2.

Applications of milling of metal components used for joint replacements are discussed in Chapter 7. Milling, however, can also be used in the

Table 6.2 Cutting speeds for milling.

Material	Cutting speed (m/min)
Stainless steels	90–500
Thermoplastics and thermosets	90–1400
Titanium alloys	40–150

production of polymer parts for tissue engineering, as described by Rouzrokh *et al.* (2010).

6.5.2 Grinding

A form of milling termed 'upmilling' has some similarity to the process of surface grinding. Grinding is a chip-removal process in which the cutting tool is an individual abrasive grain (Venkatesh and Isman 2007). The cutting points are irregularly shaped, and distributed in a random fashion along the periphery of the grinding wheel. Material is removed from the component in three stages, (i) rubbing, (ii) ploughing, and (iii) cutting. With rubbing, friction against the surface of the workpiece absorbs power without useful work being performed. In ploughing, the abrasive grit plastically creates a groove, alongside which are left particles of metal. As the wheel depth of cut rises, the cutting forces increase, leading to a cutting action with formation of chips ahead of the abrasive grits.

6.5.3 Turning

Parts made by turning are produced by rotation of the workpiece on a lathe with a cutting tool fed in an appropriate direction. A wide variety of shapes can be obtained, as noted in Figure 6.4.

The shapes may be straight, curved, flat (called 'facing'), or grooved. Single-point cutting tools are usually employed at speeds of 0.15 to 4 m/s. 'Roughing' is obtained for depths of cut more than 0.5 mm with feed rates of 0.2 to 2 mm/rev, a tolerance of about 0.13 mm being typical. Smaller depths and feeds are needed for finishing, the tolerance then being about 0.05 to 0.13 mm. The cutting tools used have to possess sufficient hardness, toughness and wear resistance and should be chemically inert, as discussed above, for milling. The turning operation can be facilitated by use of a cutting fluid rather than 'dry' machining.

Table 6.3 shows cutting speeds obtained, with materials used in joint replacements.

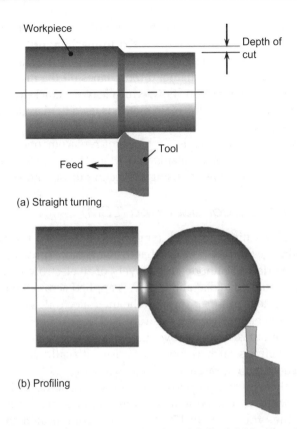

(a) Straight turning

(b) Profiling

Figure 6.4 Examples of cutting by turning. (a) Straight turning (b) Profiling

Some of these materials, like titanium alloys, can be difficult to machine by conventional methods, and alternative techniques such as electrochemical machining (ECM) may be used, as discussed in Section 6.5.4. Thermoplastics and thermosetting plastics are machined with small depths of cut, and at high speeds. Cooling by air or vapour jet or even water-soluble oils is usually

Table 6.3 Cutting speeds for turning.

Material	Cutting speed (m/min)
Stainless steels	50–300
Thermoplastics and thermosets	90–240
Titanium alloys	10–100

Source: After Kalpakjian, S. and Schmid, S.R. (2003) *Manufacturing Processes for Engineering Materials,* 4th edn, Prentice Hall. Upper Saddle River, New Jersey, USA.

recommended. Ceramics can be machined by use of ductile regime cutting. Ductile regime cutting is a technique used to machine brittle materials on the principle that all materials deform plastically provided that the scale of deformation is very small. A diamond tool planes off the material at micro-scale level, removing it by plastic flow, which leaves a crack-free finish (Ngoi and Sreejith 2000). However, the main methods used to manufacture ceramic parts for joint replacements are discussed below.

The turning of special threads for hip-joint replacements has been cited by Hoermansdoerfer (2000). Galanis and Manolakos (2009) report on the manufacturing of femoral heads using the technique of high-speed turning.

6.5.4 Electrochemical Machining (ECM)

Electrochemical machining is used to machine alloys used for parts of knee and hip prostheses. Figure 6.5 shows that the workpiece (prosthesis) and tool are made the anode and cathode, respectively, of an electrolytic cell. A potential difference, usually fixed at about 10 V, is applied across them. A suitable aqueous electrolyte is pumped through the inter-electrode gap. The cathode shape remains unchanged during electrolysis. The flow of electrolyte, whose conductivity is about 20 S/m, removes the products of machining and diminishes unwanted effects, such as those that arise with cathodic gas generation and electrical heating. The rate at which metal is then removed from the anode is approximately in inverse proportion to the distance between the electrodes. A typical gap width should be about 0.4 mm and the average current density should be of the order of 50 to 150 A/cm^2. If a complicated shape is to be

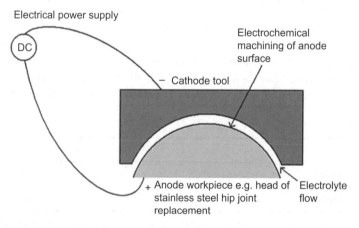

Figure 6.5 Electrochemical machining (ECM) of head of hip replacement.

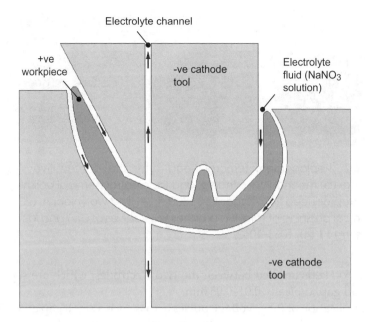

Figure 6.6   ECM of femoral component of knee replacement.

formed on a workpiece of a hard material, the complementary shape can first be produced on a cathode of softer metal, and the latter electrode is then used to electrochemically machine the workpiece. A shape, the approximate image to that of the cathode, will be reproduced on the anode. ECM can be used to shape hard metal alloys (McGeough 1974). Figure 6.6 shows the shape of a cathode tool needed to shape a knee part made of cobalt-chrome.

Zaborski *et al.* (2011) have used conventional turning, mechanical grinding, followed by a low-current density form of ECM to electropolish the heads of hip prostheses made of titanium alloy, Ti6Al4V, achieving a surface finish of 0.05–0.15 μm. (See figure 6.7.) ECM has been used to produce micro-patterns on titanium for improved osteointegration (Xiong and Yang, 2005).

6.5.5 Electrodischarge Machining (EDM)

Electrodischarge machining resembles ECM in that it employs two metal electrodes, one being the tool of a predetermined shape, and the other being the workpiece. The electrodes are immersed in a dielectric liquid such as paraffin or light oil, not an electrolyte as in ECM. A series of voltage pulses, usually of rectangular form of magnitude about 80 to 120 V and of frequency of the

Figure 6.7 Replacement femoral head after the consecutive stages of (a) turning, (b) mechanical grinding, and (c) electrochemical polishing.
Source: Reproduced from Zaborski, S., Sudzik, A. and Wołyniec, A. (2011) Electrochemical polishing of total hip prostheses. *Archives of Civil and Mechanical Engineering* **11** (4), 1053–1062.

order of 5 kHz, is applied between the two electrodes, which are separated by a small gap, typically 0.01–0.05 mm.

The application of these voltage pulses across such a small gap gives rise to electrical breakdown of the dielectric, and subsequently sparks in a localised region of the electrodes, which cause vaporization of a tiny portion of the workpiece. The cumulative effects of a succession of sparks spread over the entire workpiece leads to its erosion, or machining to a shape which is approximately complementary to that of the tool. The latter is made of electrically-conductive materials such as brass, copper or graphite. The rate of machining is independent of the hardness of the metal to be shaped. Complex shapes can be machined to fine accuracy. Van Roeckel (1992) describes using electrodischarge machining for prosthesis manufacture.

6.6 Forging

In its simplest form, 'open-die' forging consists of compressing a solid workpiece (cylindrical) between two flat dies or platens. (The technique is also called 'upsetting'.) The surfaces of the dies may be shaped so that the ends of workpieces are formed to a specific shape. Another method is 'impression die' forging. The workpiece takes the shape of the impressions of the die cavities as the dies close up on it. The radial outward flow of material as the dies meet is termed 'flash'. In 'closed-die' forging, no flash emerges. Thus a forging of required dimensions is directly obtained. Forging techniques in which the part is obtained with dimensions close to its final requirement is also termed precision or near net-shape production. Special dies are needed; metals such

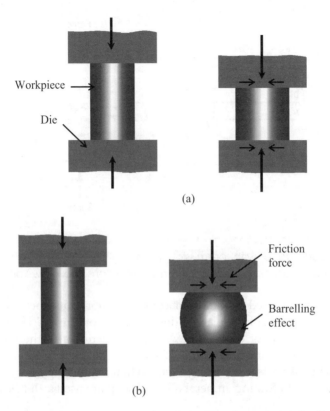

Figure 6.8   Open-die forging (upsetting) (a) with adequate lubrication (b) with inadequate lubrication.
Source: Adapted from Kalpakjian, S. and Schmid, S.R. (2003) Manufacturing Processes for Engineering Materials, 4th edn, Prentice Hall. Upper Saddle River, New Jersey, USA.

as steel and other alloys can be difficult to forge in this way. Material flow within the dies can be improved by isothermal forging in which the dies are heated to the same temperature as that of the blank. ('Hot' forging at these temperatures is the usual term. Otherwise it can be termed 'cold working' as at room temperature.) Figures 6.8 and 6.9 illustrate the main features of respectively open-die and impression die forging.

Dies should possess sufficient strength, toughness, hardness and resistance to wear in order to perform at high temperature, and to withstand mechanical and thermal shocks. Typical steel dies contain chromium, nickel, molybdenum and vanadium. Table 6.4 summarizes typical forging temperatures for

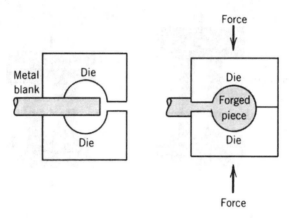

Figure 6.9   Impression die forging.
Source: Reproduced with permission from Callister, W.D. (1994) *Materials Science and Engineering,* 3rd edn, John Wiley & Sons, Inc. New York.

relevant materials used in joint replacements. Lubricants such as graphite and molybdenum disulphide are widely used in hot forging. Their purpose is to reduce the rate of cooling the workpiece, acting as a thermal barrier between the hot forging and the die. They can also affect friction and wear. The lubricant also acts as a parting agent so that the forging does not adhere to the dies. In cold forging, mineral oils and soap are used as lubricants.

6.7 Casting

6.7.1 Casting of Metals

Molten metal is poured into the cavity of a mould that has been shaped to the size and dimensions required for the implant. On solidification, the metal adopts the latter form.

Table 6.4   Hot forging temperatures (°C) for alloys used in joint replacement.

Metal alloy	°C
Alloy steels	925–1260
Titanium alloy	750–795
Refractory alloy (including chromium-based)	975–1650

Source: Adapted from Kalpakjian, S. and Schmid, S.R. (2003) Manufacturing Processes for Engineering Materials, 4th edn, Prentice Hall. Upper Saddle River, New Jersey, USA.

'Sand casting' is a common method, although not often reported for implant manufacture. Sand is packed around a pattern that bears the shape required for the part, in order to provide the 'expendable' mould. When the pattern is removed, the resulting cavity is filled with molten metal. A gating system is usually incorporated into the mould to facilitate the flow of metal. The metal is allowed to cool and solidify. It is then broken away from the sand mould and removed from the casting. The pattern for the mould can be made from wood, plastic or metal, or directly by rapid prototyping (RP), discussed in Chapter 7.

The surface finish achieved in sand casting is influenced by the coarseness of the sand from which the mould has been made. Its dimensional accuracy is generally inferior to that obtained by other methods of casting.

Die casting, a permanent mould technique, is one such alternative. Two parts of a steel mould or die are clamped together to form a cavity within which the implant metal can be formed. Molten metal is introduced into the die cavity at pressures from about 20 to 70 MPa. The method is particularly suitable for low melting-point alloys such as magnesium. On completion of solidification, the two pieces of the die are opened to yield the cast metal.

Investment (lost-wax) casting uses a pattern made from solidified wax or plastic that has a low melting temperature. The pattern is manufactured by pouring the liquid wax into a metal die made to the shape of the pattern. On its removal, the solid pattern is then dipped into a fluid slurry composed of a refractory material that coats it. When the initial coating has dried, this procedure is repeated until the required thickness of the refractory material is achieved – the 'investment'. The mould is then dried in air, and heated to a temperature of about 90 to 175 °C for 4 to 5 h. The wax pattern melts. The result is a mould, the cavity of which has the required shape and detail. Molten metal is then poured into the mould, and allowed to solidify. The mould is then broken to yield the casting. Investment casting is used to provide the high quality surface finish and good dimensional accuracy of the alloys needed for joint replacement, often without the need for subsequent surface finishing.

6.7.2 Casting of Ceramic Parts

The ceramic materials are ground or crushed into fine particles, then mixed with other constituents, which are needed to provide conditions such as binding or particle lubrication for ease of release from moulds. Ceramics are usually shaped by slip-casting, plastic forming, or pressing.

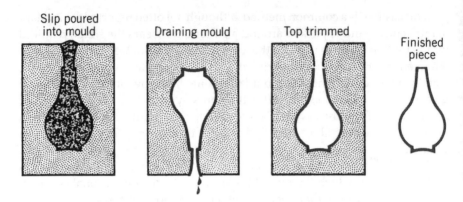

Figure 6.10   Slip casting of ceramic.
Source: Reproduced with permission from Kingery W.D. (1960) *Introduction to Ceramics,* John Wiley & Sons, Inc., New York.

6.7.2.1 Slip Casting

Figure 6.10 shows the main feature of slip casting of ceramic parts. The aqueous mixture of ceramic particles is poured into a porous, plaster of Paris mould. The mould first absorbs some of the water from the outer layers of the suspension. It is then inverted and the remaining suspension is poured out of the top of the mould. The top is trimmed. The mould is then opened.

6.7.2.2 Plastic Forming

Extrusion is a common type of plastic forming of ceramic. Figure 6.11 illustrates typical equipment. A ceramic clay mixture, usually containing about 20–30% water, is fed into a hopper, and conveyed to the melting section of the extrusion equipment. The barrel is heated, as can be the die. The barrel/screw section has three main parts, as noted in Figure 6.11. Friction and external heat soften and melt the ceramic mixture. In the mixing and pumping section, the ceramic is formed into a fluid-like substance. It is then moved into the pumping section, where the material is prepared to the required rheology for delivery to the die. The die forms the plastic mass of ceramic to its desired shape. Die design is of major significance in the extrusion of ceramics. A useful discussion is given in Ruppel (1991). The ceramic part is finally dried and fired.

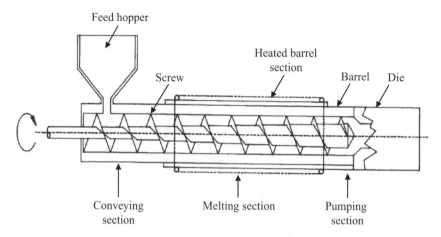

Figure 6.11 Extrusion of ceramic.
Source: Adapted from Ruppel, I. (2000) *Engineered Materials Handbook,* vol. 4: Ceramics and Glasses – Extrusion (volume chairman: Samuel J. Schneider), 2nd edn, ASM International, OH, pp. 166–180.

6.7.2.3 Pressing

The main techniques of pressing are 'dry', 'wet', and 'isostatic'. In 'dry' pressing, a punch compacts a ceramic mixture at a pressure of 35 to 200 MPa, with a highly wear resistant die, usually of hardened steel or carbide, being employed to shape the ceramic.

With 'wet' pressing, the ceramic part is produced from a mould by use of a hydraulic or mechanical press.

Ceramic parts with uniform density and grain structure and fine surface accuracy can be produced by 'isostatic' pressing. Figure 6.12 explains hot isostatic pressing (HIP). A vessel is filled with the ceramic mixture. It is then subjected to a vacuum bake, before being placed in the press container, which contains inert gas or a vitreous fluid, at 100 MPa at 1100 °C.

6.7.2.4 Injection Moulding

Another method of precisely forming ceramics is powder injection moulding. This technique is used both for metals and ceramics. For ceramic moulding, fine powders are blended with a thermoplastic polymer, such as polypropylene, which can duly be removed by a form of pyrolysis, followed by sintering

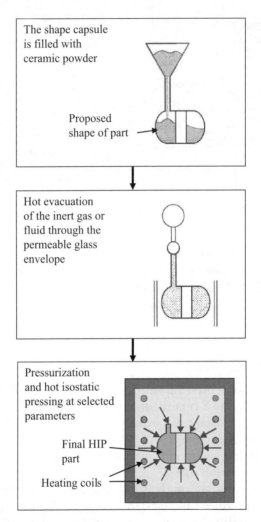

Figure 6.12   Hot isostatic pressing.
Source: Adapted from Ruppel, I. (2000) *Engineered Materials Handbook,* vol.
4: Ceramics and Glasses – Extrusion (volume chairman: Samuel J. Schneider),
2nd edn, ASM International, Novelty, OH, pp. 166–180.

(see section 6.7.2.5). Ceramic components of thickness up to 15 mm can be produced by this technique. This technique is discussed in greater detail for the processing of polymers in section 6.8.

6.7.2.5 Sintering (Firing)

The appropriate strength and hardness of the ceramic part is achieved through 'sintering'. The item is heated to a specific temperature, typically greater than 1000 °C, usually about 70–90% of its melting point in a furnace, the atmosphere for which is carefully controlled. These conditions are conductive to bonding of the oxide particles and reduction in porosity, and hence promote strength and hardness.

6.7.2.6 Shaping or Finishing of Ceramics

Processes used to shape the ceramic parts include mechanical grinding, or ultrasonic, laser or water-jet machining (the latter unconventional methods are described in McGeough 1988). Tumbling is often used to impart a final surface finish. Coating with a glaze material is used to yield a glassy finish (see section 6.10).

6.8 Manufacture of Polymer Parts

Polymer materials used in joint replacements are usually first produced in pellet or powder form, or in sheet, plates, rods or as tubes, and sometimes in liquid form. The following manufacturing processes may be considered in the manufacture of polymer components in joint replacements.

In extrusion, the polymer, in one of the above forms, is placed in a hopper. It then enters an extruder barrel, which is fitted with a screw mechanism and heating elements. A pump feeds the polymer material into the main section of the barrel, where it is melted, and then pumped into the die section. Figure 6.13 shows the main features of polymer extrusion.

On leaving the die, the polymer part is cooled in air or by passing through a water-filled vessel. The technique enables parts of complicated shape and uniform cross-section to be manufactured.

Figure 6.14 illustrates the characteristics of injection moulding. In this technique, polymer pellets or granules enter a heated cylinder in which they melt. A reciprocating screw mechanism forces the polymer material into a mould in the form of a split-die chamber. The amount of polymer injected is controlled by the specified distance travelled by the screw as it reverses when

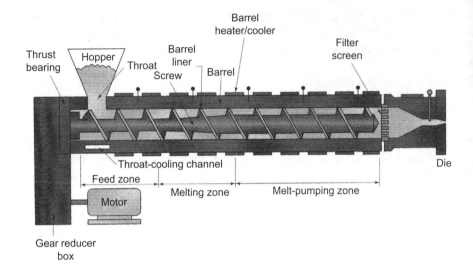

Figure 6.13   Polymer extrusion.
Source: Reproduced with permission from Mark, H.F., and Kroschwitz, J.I. (2003)
Encyclopaedia of Polymer Science and Technology, 3rd edn, John Wiley &
Sons, Inc., New York.

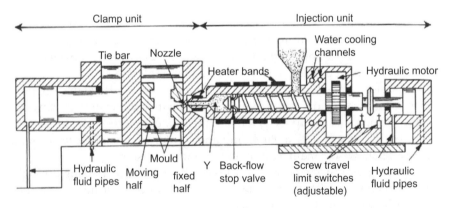

Figure 6.14   Injection moulding of plastics.
Source: Reproduced from McCrum, N.G., Buckley, C.P. and Bucknall, C.B.
(1988) *Principles of Polymer Engineering,* Oxford University Press, New York.

pressure increases at the entrance to the mould. The screw rotation is then halted; it moves forward under hydraulic pressure of 70 to 200 MPa so that the molten plastic enters the cavity of the mould. Injection moulding enables plastic parts of complex shape and high accuracy to be manufactured.

With compression moulding, a polymer, which may be in the form of a mixture of liquid resin and filler, or as a powder, is put in a sealed mould. The upper part of the die presses onto the polymer, forming it to a required shape, the dimensional accuracy of which can be superior to that produced by injection moulding.

Thermoplastics can also be cast. The monomer, catalyst and additives are heated, and the liquid mixture is placed in the mould. Polymerization proceeds, enabling the item to be produced at ambient pressure.

Thermoplastics can be formed at room temperature, known as 'cold forming', by techniques widely used for metals, including extrusion and closed-die forging, discussed earlier in this chapter. Cold-forming can yield polymer parts that have increased strength and toughness, and at lower cycle times compared to other processes of moulding.

6.9 Surface Treatment

6.9.1 Coatings

Cement-free insertion of implants is an attractive option for joint replacements. The aseptic loosening of components of the replacement associated with cemented prostheses could be reduced. The principle underlying joint replacements without cement lies primarily with promoting bone tissue growth at the interface between the prosthesis and neighbouring bone. To that end, surface coating of the prosthesis is a key feature. Coatings have a thickness of about 1–5 µm. The materials selected for coatings have to be compatible with surrounding tissue, and also resistant to corrosion, as they are in contact with body fluid. They have to adhere firmly and stably to the materials of the implant. Finally their attachment should not adversely affect the mechanical behaviour, especially fatigue resistance, of the prosthesis. The following methods of coating are used.

6.9.2 Plasma Spraying

A plasma is a partially ionized gas composed of charged particles possessing high temperature, about 10 000 K, and with high kinetic energy.

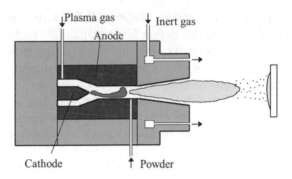

Figure 6.15   Plasma spray process.

When energy is supplied to the gas, its molecules increase in velocity. The gas temperature increases. Sufficient increase in velocity causes disintegration of the molecules into atoms, or 'dissociation'. On application of a high electric field, electrons are removed from the shells of atoms causing their 'ionization'. The consequence of this process of dissociation and ionization is the formation of the neutral plasma with its contents of an equally balanced amount of electrically positive- and negative-charged particles.

If a gas is passed through an electric arc, it can ionize on its electrical breakdown, the gas stream is then converted into a hot plasma jet, known as a 'torch', characterized by its temperature and velocity. Most plasma torches employ a direct current (DC) system, operated at atmospheric pressure, with N_2 gas mixtures being commonly used. A plasma jet is generated by the striking of a DC arc between the two electrodes. The temperature of gas flowing between the electrodes can be 6000–10 000 K. The gas is partially ionized by the electric arc thereby creating the plasma. The plasma spraying process is illustrated in Figure 6.15.

Hydroxyapatite (HA), in the form of powder, is injected through a feed port or even internally into the plasma, and is promptly melted. Its molten particles undergo acceleration towards the substrate, on which they deposit. These deposits resemble pancake-shaped lamellae or 'splats' of the order of micrometre thicknesses and lateral dimensions of several to hundreds of micrometres. The spreading is limited by the solidification process, a typical cooling rate being about 10^6 K/s.

The attraction of plasma-sprayed HA onto metal alloy implants is the formation of strong bonds with bone tissue.

Investigations continue on improvement of the mechanical properties of the HA coatings on metallic substrates, and maintenance of biocompatibility (Morks and Kobayashi 2007). Drawbacks reported with plasma sprayed HA include decomposition, and adverse effects on induced stress of rapid heating and cooling.

Plasma spraying of a porous titanium coating on a titanium (Ti6Al4V) hip stem was used in the manufacture of a prosthesis designed specifically for younger patients (Sanz-Reig *et al.* 2011). Computed tomography (CT) scans were used to establish the appropriate shape of a collared stem, which was both tapered and longer than conventional prostheses. The strain distribution for the titanium alloy stem with a cobalt-chromium-molybdenum (CoCrMo) alloy head was analysed by FEA, and experimental (photoelastic and transducer) methods; the latter was also being used to check micromotion. This implant was found to be most stable in the neutral position.

Low-energy plasma spraying has been used to deposit HA and HA/polymer (poly-ε-caprolactone (PCL)) composite coatings (Garcia-Alonso *et al.* 2011). The latter HA/polymer coating had a thickness ranging from 34 to 84 μm. It was achieved without the presence of significant residual stress, thermal degradation and coating crystallinity hitherto found with such thick HA coatings. It also provided enhanced mechanical properties compared to pure HA. The surface roughness of the coating was about 5.5 to 6.5 μm R_a. This higher surface roughness is attributed to the addition of the PCL to HA and is suggested to favour effective attachment of osteoblast-like cells to the coating surface.

A novel high-velocity oxy-fuel thermal spray method for producing hydroxyapatite (HA) coatings has been compared with other thermal spray techniques by Hasan and Stokes (2011). From experimental studies of the effect of the main process variables, oxygen, propylene, air and HA flow-rates and spray distance, they have found improvements in crystallinity and purity of the HA coatings (93.8 and 99.8%), c.f. respectively 87.6 and 99.4% for other spray methods.

6.9.3 Chemical and Physical Vapour Deposition (CVD and PVD)

Chemical vapour deposition (CVD) is a coating process that uses a reactant gas or vapour in a chamber with a heated workpiece. The gas decomposes at the heated surface of the workpiece, depositing a solid element or a compound. Material is deposited on the surface or absorbed into it. Coatings are corrosion resistant and durable. Applications include coating titanium implants with hydroxyapatite (Sato *et al.* 2007).

Plasma-assisted CVD takes place at lower substrate temperature and higher deposition rates. The reactant gas is subjected to an electrical field at frequency of 50 kHz to 13.5 MHz, or microwave frequencies. The electrical field causes collisions of the gas molecules producing ions, free radicals and other more reactive species. The presence of the plasma promotes the reaction of the chemical deposition process.

Physical vapour deposition involves vacuum processes to evaporate a coating material and deposit it onto a workpiece. Narushima *et al.* (2005) draw attention to the advantages of PVD, which allows control of the chemical composition and regulation of its crystalline and orientation properties. They report firm adhesion of a HA PVD coating to titanium. The benefit of PVD over CVD lies with its lower operating temperature. On the other hand, PVD coatings exhibit weaker bonding to the substrate.

The limitations of plasma spraying and these other methods noted above mean that other methods of coating have to be considered, some of which incorporate aspects of plasma spraying.

6.9.4 Diamond-like Carbon (DLC) Coating

These coatings, formed mainly by covalent bonding, have been described in detail by Park (2008). They exhibit an amorphous structure without grain boundaries. They contain mainly carbon (about 66%) and hydrogen (13 to 14%), with possible smaller amounts of oxygen (6%), silicon (6.5%) and nitrogen (less than 6.9%). The surfaces of diamond-like carbon coatings are smooth with average surface roughnesses of approximately 0.02 to 0.03 µm.

They can be produced by CVD, ion-beam-assisted deposition (IBAD), or plasma vapour deposition. With CVD, a vapour of carbon and hydrogen atoms is produced by thermal decomposition methods. A solid coating is produced on the substrate by condensation of the vapour.

DLC possesses high hardness, from 2000 to 3000 kg/mm^2 (Vickers scale). It also has a very low coefficient of friction. For example, the coefficient of friction for a DLC coating on CoCrMo alloy has been reported to be 0.033, compared to corresponding coefficients of 0.087 and 0.058 for respectively cast and wrought CoCrMo alloy. These combinations of great hardness and low coefficient of friction render it an attractive low wear coating material for implants. Wear rates (10^{-3} mm^3/m) of CoCrMo can be as low as 5.05 when a DLC coating is applied, compared to 6.96 for alumina coatings, and respectively, 28.7 and 20.2 for the alloy in its cast and wrought forms (Park 2008). He draws attention to the inertness of diamond-like carbon in the body. These coatings do not release metallic ions, as has been encountered with Ti6Al4V and CoCr alloys.

6.9.5 Ion Implantation

Ionized atoms of a material are bombarded against a workpiece surface in a high vacuum. They can modify the properties of the workpiece material. There is no significant dimensional change to components treated by ion implantation. Buchanan *et al.* (2004) have investigated improved wear resistance of Ti6Al4V alloy by bombardment with nitrogen ions.

6.9.6 Porous Metal Coatings

The sintering of cobalt-chrome powder or beads, or the diffusion bonding of titanium beads or wire mesh as porous coatings, in order to enhance cement fixation is becoming popular. Multiple layers of these porous materials promote bone ingrowth and facilitate stability in fixation.

The high temperatures associated with sintering of diffusion bonding can induce voids and reduce grain size thereby lowering the strength of the implant metal. Abrupt changes in dimension may also occur giving rise to increase in local stress points and consequently potential fatigue failure. For example, the porous coating of the lateral surface of a titanium stem may undergo tensile loading producing cracking. Even porous coatings of cobalt-chrome can lower the fatigue strength by 5 to 10%.

Nevertheless, porous coatings offer improved stress distribution, which should reduce the effect of such decrease in fatigue strength. The surface areas that they provide are about three to seven times that of smooth implants. This larger area may be the cause of increased release of metal ions. Abrasion of a loose implant against bone or cement can also cause ion release. In the case of titanium stems, in order to avoid the adverse effects of the release of vanadium and aluminium ions, pure titanium metal is sometimes used as the porous coating. See, for example, Hahn and Palich (2004).

6.10 Surface Finishing of Implants

6.10.1 Deburring

Removal of protuberances remaining after the machining, forging and casting of joint replacements is required before the implant is ready for use. Burrs (undesirable sharp lips or projections) can interfere with alignment of the joint replacement and can arise in all materials used in joint replacements.

In addition, to mechanical procedures such as filing and wire brushing, vibration and barrelling are commonly used. The parts to be deburred are placed in a container holding an abrasive medium such as corundum within an aqueous (or dry) medium and are then tumbled, or undergo vibration within the container; deburring is effected by the abrasive removal of the sharp edges of the component.

Alternatively, shot blasting can be used. Abrasive sand particles are projected towards the component by a blast of air. The abrasive action removes burrs, and also oxide surface films and can also clean the surface leaving a matt-like finish. A useful account of methods of deburring has been given by Chern and Dornfeld (1996) and Dornfeld *et al.* (1999).

After deburring, the implant has to be given a final surface finish. For metallic components, electropolishing is popular.

6.10.2 Electropolishing

Anodically polarized parts to be polished are placed in a cathodically polarized vessel containing an aqueous electrolyte such as orthophosphoric acid, a small voltage of 3–6 V being applied between the electrodes. Preferential electrolytic dissolution of the peaks on the surface of the component leaves a smooth and polished finish. Unlike ECM ($100 \, A/cm^2$) discussed earlier, in electropolishing low current densities ($10^{-2} \, A/cm^2$) are used with the removal surface irregularities as small as $10^{-2} \, \mu m$.

Zaborski *et al.* (2011) have described a sequence of computer-aided numerically controlled (CNC) turning, followed by mechanical grinding, mechanical and electrochemical polishing of titanium alloys (Ti6Al4V) used in the production of the head component of hip replacements. Grinding was found to leave a roughness of about $0.3 \, \mu m \, R_a$, with residual surface cracks. Subsequent mechanical polishing by lapping-based techniques still left microcracks, although a surface roughness of $0.116 \, \mu m \, R_a$ was achieved. Their presence was considered to be conducive to possible stress-corrosion of the implant. Electrochemical polishing was concluded to be a finishing procedure that could overcome these difficulties. To that end an electrolyte composed of hydrochloric and perchloric acids with chromium oxide was used to achieve an electropolished surface finish of 0.05 to $0.15 \, \mu m \, R_a$.

6.10.3 Mechanical Polishing

For mechanical polishing, discs or belts are coated with abrasive aluminium oxide or diamond powder. The abrasive and frictional action removes small-scale surface irregularities leaving a lustrous finish. Polishing can subsequently undergo buffing by the effect of very fine abrasives coated on discs made from cloth or hide.

Charlton and Blunt (2008) report on polishing of knee joints made of cobalt-chrome alloy. Its surface was first progressively hand polished with diamond cloths to reduce surface finish from 178 nm to between 9.7 nm and 2.8 nm R_a. They claim that this procedure should produce knee joints to the required form and nano-scale surface topography, extending their lifetime.

6.10.4 Lapping

Lapping is a process by which material is precisely removed from a workpiece to produce a desired dimension, surface finish, or shape. Flat or cylindrical parts of the implant can be finished by the effects of abrasive particles on a lapping material usually made of leather or cloth. Copper lapping plates can also be used to yield a high-quality surface finish (Zaborski *et al.* 2011).

6.11 Manufacture of Joint Replacements

Precise details of the industrial manufacturing processes used to make joint replacements can be difficult to procure. However, some examples of the machining, casting, forging and other processes and techniques discussed in this chapter can be presented. For instance, titanium femoral stems are known to be machined with multiple and precision tools being used to produce the general and final shape. Typical times can be about 30 min for a single part. Robotic-controlled sanding belts are used to remove undesired surface marks, and to yield the required surface finish. The replacement part subsequently undergoes simulated joint movement by wear-testing in artificial joint fluid. Thermal spraying is used to facilitate the adherence of bone tissue to the part. The dimensions of the part are checked with an optical comparator by projection of an image-coded outline against a template of the required size.

The complexity of shape of other metallic parts such as femoral components and the tibial tray means that computer-aided design (CAD) is first used, prior to manufacture by investment casting. Undesired impurities and residual stresses left within the casting are removed by techniques such as hot isostatic pressing. Employment of high pressures and temperatures just below the melting point of the alloy enable modifications of its microstructure with a decrease in grain size, which endows the alloy with its required mechanical properties.

When trabecular metal and other porous structures are to be used for tibial trays and patella attachments, they can be made by methods such as powder sintering, plasma spraying, low-temperature diffusion bonding and selective laser melting (SLM). With (SLM), a high-powered *neodymium*-doped yttrium *aluminium* garnet (Nd:YAG) laser is used to melt a metal powder, typically titanium, such that the porosity of the product is close to that of new bone (see, for example, Vandenbrouke and Kruth 2006 and Warnke *et al.* 2009).

The wear-resistance of the metallic replacement joints can be enhanced by application of a ceramic coating to the alloy. In total knee replacement production, physical (PVD) and chemical (CVD) vapour deposition is used. Other methods including nitrogen diffusion hardening and nitrogen ion implantation are also employed. These coating techniques are costly, requiring several days, with high energy consumption.

After the joint replacement has been ceramic coated, it has to be surface finished, by techniques such as polishing. It is endowed with a mirror finish and smoothness necessary for its articulation as a joint.

Compression moulding is used to press UHMWPE sheets in a single operation in a computer-controlled heating and hydraulic loading system, in a clean room. This procedure can take about 24 hours to obtain the microstructure and mechanical properties required. Ram extrusion can also be employed to manufacture bulk UHMWPE (Kurtz 2004).

Direct compression moulding is well used. It provides a means of direct manufacture of prosthetic implants with a surface finish that is highly polished. Thermoplastic manufacturing methods such as injection moulding and screw extrusion are not suitable for UHMWPE, as the material does not flow like lower molecular-weight polyethylenes when it is raised above its melt temperature.

References

Archard, J.F. (1953) Contact and rubbing of flat surfaces. *Journal of Applied Physics* **24**, 981–989.

Buchanan, R.A., Rigney, Jr., E.D. and Williams, J.M. (2004) Ion implantation of surgical Ti-6Al-4V for improved resistance to wear-accelerated corrosion. *Journal of Biomedical Materials Research* **21** (3), 355–366.

Callister, W.D. (1994) *Materials Science and Engineering – An Introduction*, 3rd edn, John Wiley & Sons, Inc., New York.

Charlton, P. and Blunt, L. (2008) Surface and form metrology of polished 'free-form', biological surfaces. *WEAR* **264**, 394–399.

Chern, G.L. and Dornfeld, D.A. (1996) Burr/breakout development and experimental verification. *Transactions of the American Society of Mechanical Engineers, Journal of Engineering Materials Technology* **118** (2), 201–206.

Dornfeld, D.A., Kim, J.S., Dechow, H. *et al.* (1999) Drilling Burr formation in titanium alloy, Ti-6Al-4V. *Annals of the International Academy for Production Engineering* **48** (1), 73–76.

Galanis, N.I. and Manolakos, D.E. (2009) Surface roughness of manufactured femoral heads with high speed turning. *International Journal of Machining and Machinability of Materials* **5** (4), 371–382.

Garcia-Alonso, D., Parco, M., Stokes, J. and Looney, L. (2011) Statistical design of experiment of low energy plasma spray deposition of hydroxyapatite/poly-ε-caprolactone biocomposite coatings. *Journal of Thermal Spray Technology* **21** (1), 132–143.

German, R.M. (1994) *Powder Metallurgy Science,* Metal Powder Industries Federation. Princeton, NJ, U.S.A.

Hahn, H. and Palich, W. (2004) Preliminary evaluation of porous metal surfaced titanium for orthopedic implants. *Journal of Biomedical Materials Research* **4** (4), 571–577.

Hasan, S. and Stokes, J. (2011) Design of experiment analysis of the Sulzer Metco DJ high velocity oxy-fuel coating of hydroxyapatite for orthopedic applications. *Journal of Thermal Spray Technology* **20** (1–2), 186–194.

Hoermansdoerfer, G. (2000) Hip joint glenoid cavity with a special thread. US Patent no. 6 146 425, 14 November 2000.

Kalpakjian, S. and Schmid, S.R. (2003) *Manufacturing Processes for Engineering Materials,* 4th edn. Prentice Hall. Upper Saddle River, New Jersey, USA.

Kingery, W.D. (1960) *Introduction to Ceramics,* John Wiley & Sons, Inc., New York.

Kruth, J.P. and Vandenbroucke, B. (2006) Selective laser melting of biocompatible metals for rapid manufacturing of medical parts. *Rapid Prototyping Journal* **13** (4), 196–203.

Kurtz, S.M. (2004) *The UHMWPE Handbook: Ultra-High Molecular Weight Polyethylene in Total Joint Replacement,* Elsevier Academic Press, London.

Mark, H.F. and Kroschwitz, J.I. (2003) *Encyclopaedia of Polymer Science and Technology,* 3rd edn, John Wiley & Sons, Ltd, Chichester.

McCrum, N.G., Buckley, C.P. and Bucknall, C.B. (1988) *Principles of Polymer Engineering,* Oxford University Press, New York.

McGeough, J.A. (1974) *Principles of Electrochemical Machining,* Chapman & Hall, London.

McGeough, J.A. (1988) *Advanced Methods of Machining,* Chapman & Hall, London.

Morks, M.F. and Kobayashi, A. (2007) Microstructure and mechanical properties of HA/ZrO_2 coatings by gas tunnel plasma spraying. *Transactions of Japan Welding Research Institute* **36** (1), 47–51.

Narushima, T., Ueda, K., Goto, T. *et al.* (2005) Preparation of calcium phosphate films by radiofrequency magnetron sputtering. *Mater Trans* **46**, 2246–2252.

Ngoi, B.K.A. and Sreejith, P.S. (2000) Ductile regime finish machining – a review. *International Journal of Advanced Manufacturing Technology* **16**, 547–550.

Park, J.B. (2008) *Bioceramics: Properties, Characterizations and Applications*, Springer Science and Business Media, LLC, New York.

Park, J.B. and Bronzino, J.D. (2002) *Biomaterials Principles and Applications*, CRC Press, LLC, Boca Raton, FL.

Rouzrokh, A., Yi-Hsuan Wei, C., Erkorkmaz, K. and Pilliar, R.M. (2010) Machining porous calcium polyphosphate implants for tissue engineering applications. *International Journal of Automation Technology* **4** (3), 291–292.

Ruppel, I. (1991) Extrusion, in *Engineered Materials Handbook*, ASM International **4**, 166–172. Novelty, OH. pp. 166–180.

Sanz-Reig, J., Lizaur-Utrilla, A., Llamas-Merino, I. and Lopez-Prats, F. (2011) Cementless total hip arthroplasty using titanium, plasma-sprayed implants: a study with 10 to 15 years of follow-up. *Journal of Orthopaedic Surgery* **19** (2), 169–173.

Sato, M., Tu, R., Goto, T. *et al.* (2007) Hydroxyapatite formation on calcium phosphate coated titanium. *Materials Science Forum* **561–565**, 1513–1516.

Skalski, K. and Grygoruk, R. (2004) Manufacture process of modular radial head mobile endoprosthesis and its instruments. Proceedings of IIIth International Conference on Advances in Production Engineering, Part III, pp. 41–46. Warsaw, Poland.

Van Roekel, N.B. (1992) Prosthesis fabrication using electrical discharge machining. *Int J Oral Maxillofac Implants* **7**, 56–61.

Vandenbroucke, B. and Kruth, J.-P. (2007) Selective laser melting of biocompatible metals for rapid manufacturing of medical parts. *Rapid Prototyping Journal* **13** (4), 196–203

Venkatesh, V.C. and Izman, S. (2007) *Precision Engineering*, Tata McGraw-Hill Publishing Company Ltd, New Delhi, India.

Warnke, P.H., Douglas, T., Wollny, P. *et al.* (2009) Rapid prototyping: orous titanium alloy scaffolds produced by selective laser melting for bone tissue engineering. *Tissue Engineering Part C: Methods* **15** (2), 115–124.

Xiong, L. and Yang, L. (2005) Electrochemical micromachining of titanium surfaces for biomedical applications. *Journal of Materials Processing Technology* **169** (2005), 173–178.

Zaborski, S., Sudzik, A. and Wołyniec, A. (2011) Electrochemical polishing of total hip prostheses. *Archives of Civil and Mechanical Engineering* **11** (4), 1053–1062.

7

Computer-Aided Engineering in Joint Replacements

7.1 Introduction

Computer-based technology is aimed at optimizing surgery needed for joint replacement. Early applications of computer-aided joint replacement were envisaged as lying in medical information retrieval systems, prosthesis design, computer graphics and rehabilitation, for example in gait analysis. Picard *et al.* (2004) reviewed these developments. The manufacturing processes discussed in the previous chapter are being increasingly backed up by computer-aided procedures. To that end, conceptual designs of new kinds of joint replacement can be first evaluated. Computer-aided design (CAD) is used to define the physical and functional characteristics of the joint replacement. It can be iterative until an acceptable design is reached and can be used to evaluate eventual performance. The data gathered can be transferred for computer-controlled manufacture of the joint replacement. The principles of computer-aided design systems and the architecture involving hardware, software, and data collection are described by McMahon and Browne (1995).

The Engineering of Human Joint Replacements, First Edition. J.A. McGeough.
© 2013 John Wiley & Sons, Ltd. Published 2013 by John Wiley & Sons, Ltd.

7.2 Reverse Engineering

The concept of 'reverse engineering' is applied in the design of joint replacements (Raja and Fernandes 2008). Reverse engineering is concerned with the construction of a digital CAD model of a real object such as a knee or hip joint. By X-rays, CT, or MRI scans, data are acquired on the internal and external geometry of the joint, and on its density and composition. These data are captured in a format described as 'point cloud', that is, in terms of a large number of x, y, and z coordinates. The accuracy of the captured shape depends on the density of the points in the cloud and consequently upon the scanning method used. Each orientation used in scanning results in a point cloud; there are usually several point clouds for an object and they have to be merged into a single global set in a procedure termed 'registration'. The points are joined to form a polygon 'mesh', to produce a set of triangular elements representing the outer and inner surfaces of the object – that is, the human joint. The triangulation is usually done using Delaunay's principle, which tends to optimize the shape of the triangles. The triangular mesh that represents the outer surfaces of the joint is then divided into segments or patches. By subdividing it into patches, it makes it possible to represent each patch either as a Bézier or B-spline surface. Segmentation is based on the geometrical properties of the triangular elements, for example, magnitude of the normals to the triangular elements and/or the principal curvatures at each node. Finally, a solid or surface model of the joint is obtained and stored in B-rep format. The model can be subsequently processed to generate an approximate representation of the joint in STereo Lithography (STL) format for application in rapid prototyping, as described later. These procedures are shown in Figure 7.1.

These computer-aided methods may now be further illustrated in terms of an example of a human joint. To that end, the knee joint is discussed here.

The geometry of the knee joint is obtained from its basic 'parameter' entities that make up its shape, such as straight lines, circles, conics (for example parabolae and ellipses). The boundaries of the joint are defined by a technique termed 'boundary representation' (B-rep). The coordinate points that define the shape give rise to curves and surfaces. Curve-generating functions, 'Bézier' or 'B-spline' methods are used to smooth out the data points. These curves usually consist of several segments; B-splines are generally used because they ensure slope continuity between neighbouring elements. It is also possible to obtain continuity with Bézier curves but this is more cumbersome. Hence, they are usually avoided unless the geometry is straightforward which is not the case in joint anatomy. The curves are usually

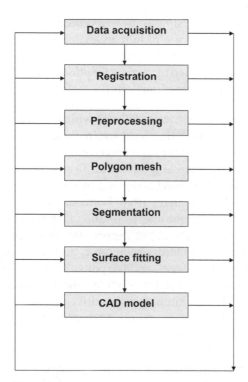

Figure 7.1   Procedure for reverse engineering.
Source: Reproduced with permission from Sri Hinduja

three-dimensional (3D) representations of the knee joint. From this information, its geometric characteristics can be represented through solid modelling.

7.3 Solid Modelling

Solid modelling is concerned with the mathematical and computer modelling
of three-dimensional solids and lies at the heart of computer aided design. For
that purpose, solid joints can be represented by 'wireframe', 'surface' or 'solid'
models. In a wireframe model, the object is defined by its bounding lines and
end-points, that is, by the equations that represent the boundary elements
(for example, lines or arcs) and the coordinates of the end points. It can
be regarded as a collection of segments that represent the edges of an object.
Wireframes can be two- (2D) or three-dimensional (3D). Owing to deficiencies
in wireframe models, such as ambiguity, they are seldom used in modelling

of human joints. Solid modelling is preferred because it represents the object as a closed volume. In the case of human anatomy, a solid model will have one or more of its surfaces as a B-spline or Bézier surface. It is also possible that the entire joint is bounded by Bézier/B-spline surfaces. Most CAD/CAM systems provide the following functions to create the solid model.

- 'Instancing and parametric modelling', which is a direct way of defining a new object from an existing one by transformation. For example, equal scaling of a cube in all three directions creates another cube but differential scaling produces a rectangular solid. To avoid distortion or confusion concerning shape, the object is defined by parameters.
- 'Sweeping'. A closed planar surface is swept either translationally along a curve, usually a straight line, or rotational about an axis. If the planar surface is not closed, the object degenerates to a curve, and the swept object is a surface. The terms 'generator' and 'director' denote respectively the curves or surfaces, and the path along which they are swept.
- 'Primitives'. All CAD systems provide functions to create some of the most common shapes, known as 'primitives' or building blocks – these primitives are a block, cone, cylinder, torus and triangular wedge.
- Boolean operations. Most joints are more complex than primitives. The latter can be combined together by used of Boolean operations, namely (i) 'union': uniting two bodies to form a third; (ii) 'difference': subtracting one body from another to form a third; (iii) 'intersection': taking two bodies and constructing a third representing the material common to both; and (iv) 'gluing': joining pairs of bodies along coincident edges or surfaces. The composition of a complex object by use of Boolean operations is usually represented graphically by use of a binary tree. Every leaf of the tree is a primitive that has been positioned correctly by a transformation, as defined above. Each nonterminal node is the result of a Boolean operation or translation. The root represents the final object. The construction of a binary tree representation of a solid object created by Boolean operations is presented in Figure 7.2.
- 'Skinning' creates a face that covers several prespecified sequences of cross-sectional shape, usually defined by the wireframe of other solid model. If the two end faces are not added to the skin face the resulting object is a surface and not a solid.
- 'Blending' is used to achieve a smooth transition between two faces, resulting in a continuity of slope.
- 'Shelling' offsets the faces of a body by a certain amount, and then subtracted from the original to create a thin-walled solid. An example of shelling is given in Figure 7.3.

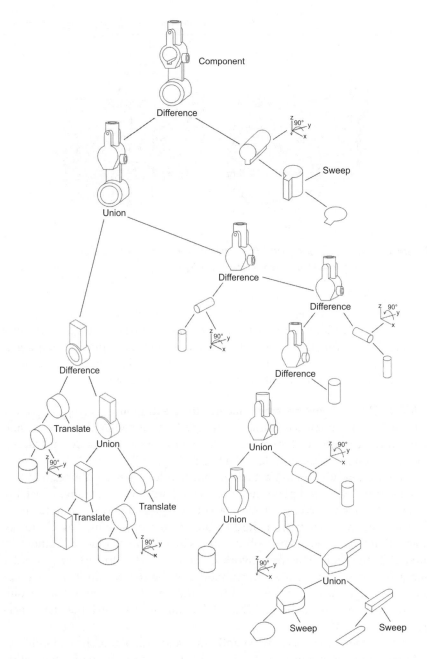

Figure 7.2 Binary tree representation of solid object created by Boolean operations.
Source: Reproduced by permission of Sri Hinduja.

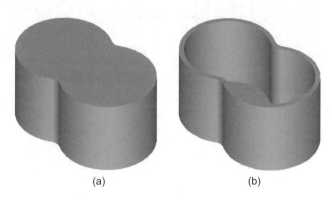

Figure 7.3   (a) solid body and (b) shelling applied to solid body.
Source: Reproduced by permission of Sri Hinduja.

- 'Local operations' are functions available to modify the geometry of a face. These functions are particularly useful in the rapid prototyping of joint prostheses when the design has to be modified to take account of special needs of a patient or feedback from the surgeon.

Most CAD systems provide a facility for creating and storing frequently used shapes in a database. These shapes can be accessed by the designer and used in the creation of the joint replacement prosthesis. The term 'feature-based modelling' is used to describe this facility.

In computer-aided manufacturing, solid modelling is the foundation that enables simulation and planning of processes, such as machining, which are to be undertaken. A comprehensive account is given by Mortenson (2006).

In summary, therefore, from the CT scans, the femur is defined by 'clouds' of points that are connected by 'spline' curves in order to produce a 'framework model' of the joint. This framework is then subjected to further computer analysis in order to yield a surface model of the upper and lower parts of the knee. Figure 7.4 illustrates (a) a three-dimensional image of a knee joint and (b) a three-dimensional CAD model of a femur to match the anatomical geometry of a patient.

Finite element analysis may be undertaken to investigate the suitability of a design, for example by calculation of the bone stress-strain characteristics. After FEA, the design can be modified iteratively until a satisfactory solution is obtained. The significance of FEA in the engineering of human joint replacements is discussed in the next section.

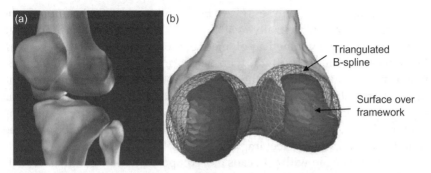

Figure 7.4   (a) A three-dimensional image of a knee joint, and (b) a 3D CAD model of a femur to match the anatomical geometry of a patient.
Source: Based on Sandholm *et al.* (2011).

7.4 Finite Element Analysis (FEA)

Finite element analysis (FEA) seeks to establish an approximate solution to a physical problem that is defined within a finite region. Usually the quantity to be found varies continuously over the region. The approximate solution determined by FEA does not have the same extent of continuity. The solution is derived by dividing up the region into a number of simple shapes, the 'finite elements'. These are located by the position of points in the region termed 'nodes'. Each element is rigidly joined to its neighbouring elements at the common nodes; assumptions are made as to the manner in which the displacement (deflection) varies within the element. Usually a linear variation is assumed. This variation ensures that there is no gap or overlap between edges of neighbouring elements.

Assumption of the linear variation of the deflection within the element, together with the Young's modulus and Poisson's ratio, defines the properties of the element. Once the properties of all the elements are calculated, they are assembled to give the global stiffness matrix for the joint. This matrix is in fact a set of linear simultaneous equations, which, when solved, gives the deflections at each node.

The accuracy of the solution depends on the density of the finite element mesh and the type of element used. If linear elements are used, then it may be necessary to use a very fine mesh. In the design of joint replacements, FEA can be used to yield two- and three-dimensional representations of models that may be used, and their performance (Phillips *et al.* (2008)). Some examples may illustrate the usefulness of FEA.

Using finite element analyses (FEA), Bougherara *et al.* (2009) compared the performance of a commercially available 316L stainless steel knee implant, with a hybrid that combined a polymer-composite (termed CF/PA-12) with the stainless steel alloy. The hybrid consisted of a layer of the polymer composite that was about one-half of the thickness of the original femoral implant, the final assembly geometry being the same as that of the original metal implant. The aim was to use the hybrid in order to alleviate stress-shielding and bone loss by transferring more load to the femur than could be expected from the solely stainless-steel implant.

The analysis began with CT scans of a composite femur and tibia. A CAD model was generated for both the cancellous and cortical geometries of the femur and tibia. Based on a commercially available press-fit condylar system, a CAD model for the proposed TKR implant was next produced.

The assembly model for FEA of the implant and its insertion into the bone was undertaken by appropriate 3D-modelling software into its assembly window. The assembly comprised the implant, distal end of the femur and proximal end of the tibia. The model was then exported for simulation. The FEA mesh needed was generated based on elements sized at 0.5 mm. Figure 7.5 shows the meshed assembly of the bone and the knee implant.

From the available data on the mechanical and other physical properties, the simulation consisted of applying typical axial loading and restraints (2100 N at the proximal – most end of the femur) which is equivalent to three times a body weight of 70 kg. The data on which they based their FEA calculations for bone are provided in Table 7.1.

The FE model was then used to generate the stress distribution contours of the 316L stainless steel and the hybrid joint replacement shown in Figure 7.6. The hybrid is noted to give a higher minimum and lower maximum stress compared to the stainless steel. From the FEA the maximum stresses generated in the UHMWPE tibial plate occur at 20 MPa, with loading have little effect over most of the layer (Figure 7.7). The extent to which finite element analysis is now being used in the engineering of joint replacements is exemplified by the cases now discussed.

Simulation, including FEA methods for assessing tribological performance of hip joint prostheses, has been cited by Fisher *et al.* (2011). They refer to a wide range of conditions, including femoral head damage and degradation of polyethylene, gait, patient type, and position of the prosthesis.

Finite element analysis, based on Abaqus software of cartilage and lateral damage in hip joints with abnormal geometry, known as femoroacetabular pincer impingement (FAI), has been performed by Hellwig *et al.* (2011). They

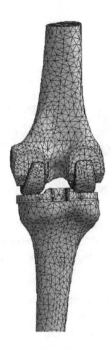

Figure 7.5   Assembly of the knee implant from finite element analysis.
Source: Reproduced with permission from Bougherara, H., Mahboob, Z., Miric, M. and Youssef, M. (2009) Finite element investigation of hybrid and conventional knee implants. *International Journal of Engineering* **3** (3), 257–266.

Table 7.1   Material properties of simulated cortical and cancellous bone

	Simulated cortical bone	Simulated cancellous bone	
		Solid	Cellular
Density (kg/m^3)	1640	270	320
Compressive strength (MPa)	157	6.0	5.4
Compressive modulus (GPa)	16.7	0.155	0.137
Tensile strength (MPa)	106	n/a	n/a
Tensile modulus (GPa)	16.0	n/a	n/a

Source: After Bougherara *et al.* (2009).

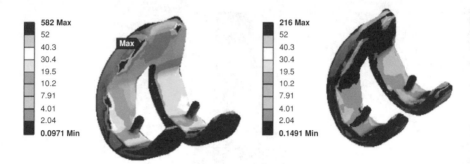

Figure 7.6   Stress distribution (MPa) in the implant (left), and the hybrid implant (right).
Source: Reproduced with permission from Bougherara, H., Mahboob, Z., Miric, M. and Youssef, M. (2009) Finite element investigation of hybrid and conventional knee implants. *International Journal of Engineering* **3** (3), 257–266.

constructed a three-dimensional solid model of a hip joint. Conditions were predicted for collision of the femoral neck against the acetabular rim causing pin impingement. (FAI can cause osteoarthritis.) It has also been studied by Cobb *et al.* (2011) who use CT-methods to quantify displacement).

Agarwal *et al.* (2009) used Abaqus to build biomechanical finite element models to investigate the depths of bone formation needed to augment defective bone stock in order to stabilize the tibial tray used in TKR. Modelling of the human knee meniscus by finite element methods has been discussed by McGeough *et al.* (2003).

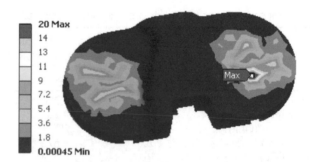

Figure 7.7   Stress distribution (MPa) in UHMWPE tibial plate.
Source: Reproduced with permission from Bougherara, H., Mahboob, Z., Miric, M. and Youssef, M. (2009) Finite element investigation of hybrid and conventional knee implants. *International Journal of Engineering* **3** (3), 257–266.

Wang and Wu (2009) used FEA to show that cyclic peak stresses greater than 15 MPa during walking can cause fatigue wear of polyethylene inserts, of linear elastic modulus and yield stress of respectively 600 MPa and 10.8 MPa, used in TKR.

High flexion activities, especially in the Middle East and Asia, make specific demands on design of TKR. Greenwald *et al.* (2009) used FEA to determine peak contact stresses on tibial inserts in designing TKR for these groups. Three-dimensional FEA was also used to evaluate the stress on polymer inserts used in designs of TKR.

7.5 Rapid Prototyping (RP) in Joint Replacement Manufacture

Rapid prototyping (RP) is used in conjunction with computer-aided design to produce a three-dimensional physically solid model of a proposed implant. The process, also called three-dimensional (3D) printing allows models to be produced in time scales of the order of hours or days in comparison with weeks needed by more established methods of manufacture. It also provides a means of manufacture of a model implant that is an accurate representation of that needed to describe a specific joint anatomy. An accurate fit may thereby be obtained with the inherent bone structure still retained. The extent of resection of healthy bone and tissue is reduced with custom-designed implants. The risk of aseptic loosening is also lessened, as the accuracy of congruency of joints should be higher than that obtained with conventionally sized prostheses. The RP procedure begins with acquisition of images from CT or MRI scans of the joints. Other image methods such as ultrasound may also be used. The images are then reverse engineered, with the external surfaces being represented by B-spline/Bézier surfaces. These surfaces are then converted into STL format, which, as mentioned before, is an unstructured triangulated representation of the image. All RP machines accept input data only in STL format.

Stereolithography apparatus (SLA) is commonly employed. Its components comprise the software embodying the design, a tank containing liquid polymer, a table incorporating vertical movement, and a laser, usually the helium-cadmium type, used to cure the polymer. Figure 7.8 illustrates the features of SLA. Under the direction of the CAD SLA the laser, focused to about 0.2 mm diameter, is directed down onto the liquid polymer precisely scanning its surface layers, typically of depth 0.12 to 0.5 mm at speeds from 0.05 to 500 mm/s. The scan is performed on a series of hatchings longitudinally, laterally, then at 45° in both directions.

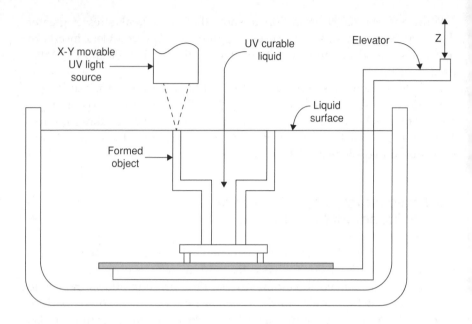

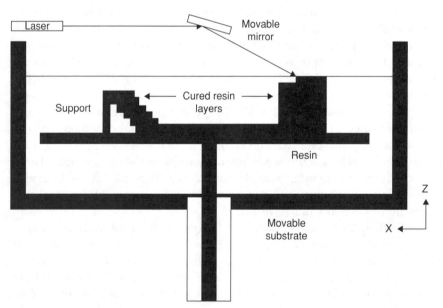

Figure 7.8 Stereolithography apparatus (SLA).

The polymer on which the laser is focused contains two basic components. One is a photoinitiator which absorbs the laser energy freeing a reactive radical species which in turn initiate polymerization. The other is photoresin containing monomers and oligons. They form a solid polymer upon exposure to the free radicals.

On completion of the first laser scan, the thin layer, about 0.125 mm thick, of cured polymer is formed on the table. The table then descends in steps of 0.125 mm for formation of the next layer. Each layer is bonded to that underneath so that an integrated structure is formed.

The component, the accuracy of which is about 0.125 mm, is then removed from the tank, and cured in an ultraviolet chamber, in order to solidify residual liquid left in the intersects during scanning.

The model of the implant so produced from the SLA should match the geometry of the individual joint, serving as a template for its manufacture. Its STL files can be used either to manufacture the prosthesis or part of it directly, to reverse engineer a wax pattern for production of a mould for casting for the implant.

Selective laser sintering (SLS) is a similar process, except that the laser beam scans layers of sintered powder, which are duly solidified into the shape required. Fused-deposition modelling (FDM) is a third type of RP. A thermoplastic filament (or metal wire) is unwound from a coil and fed into a heated extrusion die. The filament material is deposited in a thin layer, about 0.04 mm thick, upon a fixtureless foundation. The extrusion head follows a path designated by the CAD file. The part is duly manufactured in an additive process. Lee *et al.* (2009) have described FDM in the manufacture of a femoral component for a knee replacement. Figure 7.9 illustrates how the laser is used in construction of a three-dimensional solid model of a femur derived from CAD reconstruction from X-ray or CT images. Horáček *et al.* (2010) also used FDM describing how a wax pattern was employed for direct precision casting of knee joint replacement parts.

Figure 7.10 shows the solid model made in anticipation of the FDM process being used. Figure 7.11 presents the rapid prototyped models of the femoral component made by FDM.

In addition to these applications in joint replacement, rapid prototyping principles are used in the reproduction or repair of damaged tissue, such as bone and cartilage. This technology is known as tissue engineering. A three-dimensional model of the cellular structure of the tissue is constructed by computer-aided design (CAD) methods. These data are next processed by appropriate software, as discussed above for RP. A solid three-dimensional scaffold is then produced with a defined internal and external structure

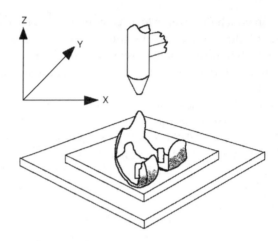

Figure 7.9   Schematic illustration of a three-dimensional image of knee femur.

that replicates that of the original cell structure. Fresh cells then attach to the scaffold. The materials from which the scaffold is made range from polymers, collagen hydrogels and ceramics to metals (Jaramillo-Botero *et al.* (2010)). When biodegradable polymer is used to form the scaffold on which fresh cells adhere, it will gradually disintegrate, being replaced by a cellular structure free of defects or damage. Carbon nanotubes have also been used as scaffold materials on which bone has been regrown, by attachment of bone cells. Scaffolds constructed with nanofibres are particularly effective.

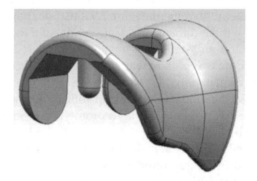

Figure 7.10   Solid model of the femoral component.
Source: Reproduced from Lee, J-N., Chen, H-S., Luo, C-W. and Chang, K-Y. (2009) Rapid prototyping and multi-axis NC machining for the femoral component of knee prothesis, *Life Sciences Journal* **6** (3), 73–77.

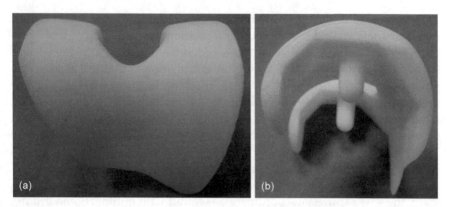

Figure 7.11 Rapid prototyping of femoral component by FDM: (a) front view and (b) side view.
Source: Reproduced from Lee, J-N., Chen, H-S., Luo, C-W. and Chang, K-Y. (2009) Rapid prototyping and multi-axis NC machining for the femoral component of knee prothesis, *Life Sciences Journal* **6** (3), 73–77.

Owing to the fibrous character of most tissues, cells attach to the fibres and grow along them.

Three-dimensional scaffolds composed of hydroxyapatite (HA)/poly (ε-caprolactone) (PCL) required for tissue engineering application have been manufactured by selective laser sintering (SLS) (Eosoly *et al.* 2009). By control of the SLS operating conditions, such as the laser power (25 W), spot size (410 μm), scanning speed (1800 mm/s), and HA/PCL particle size (125 μm), they produced a scaffold with required dimensions such as strut and pore size of respectively 0.6 mm and 1.2 mm, and with specific mechanical properties for example yield strength (0.1 to 0.6 MPa). The prospects of osteoregeneration instead of bone allografts by tissue engineering techniques have prompted Smith *et al.* (2011) to investigate the use of tantalum trabecular metal in supporting skeletal cell growth.

7.6 Computer-Aided Manufacture

The RP model of the joint replacement can be viewed by the surgeon and modified accordingly. On acceptance, its manufacture can proceed.

Computer-aided design data have to be transferred to enable machining to be performed. A brief summary of the steps followed may be useful.

This transfer is achieved through computer numerically controlled (CNC) machine tools that depend upon information contained in a part program generated by specialist computer-aided manufacture software such as PathTrace and DepoCAM. The digital information contained in this program describes the movement of the machine tool in order to machine the joint. This digital information is converted into voltage signals, which are subsequently supplied to the activation devices and servo motors, putting the procedure under 'closed-loop control'. See for example Tlusty (1999).

The machine cutting action can take place over two, usually orthogonal, axes, as for example with a lathe or in a three-dimensional movement, normally the x, y, z Cartesian coordinates, as found with milling. Some drilling machines also work on this principle. Four- and five-axis movement can also be obtained, based respectively for example on three linear and one rotational axes, or three linear and two rotational axes: employed generally on milling machines. The movements can be 'point-to-point' in which the cutting tool moves from one point to another and performs operations which usually require movement along one axis. Point-to-point controllers are extensively used where operations such as drilling, reaming and threading are required. The other type of controller is of the contouring type wherein the movements of the machine are controlled in more than one direction simultaneously. For machining complex shapes containing spline curves, a three- to five-axes controller would be required. The velocity with which the tool moves is specified in the part program and is referred to as 'feed rate'. It is specified in terms of either a linear distance per revolution of the spindle (lathe operations) or linear distance per unit time (milling).

The different stages involved in obtaining the part program that controls the movement of the machine tool are shown Figure 7.12. The CAD system uses the geometry of the joint and the cutter to generate the tool paths. These tool paths contain the x-, y- and z-coordinates of the cutter as well as its inclinations to the axes. At this stage, the tool planning module ignores any information, (such as feed rate and spindle speed) related to the machine tool. Once the tool paths have been obtained, they are written to a file in an ISO format and the file is referred to as the CL data (centre line of the cutter) file. This file is then post processed, the purpose of which is to convert the tool paths and the machine related data (such as spindle speeds, feed rates, etc) into a word address format (also referred to as G and M format). Whilst most of the most of the G and M codes have been standardized, some may be specific to the controller of the CNC machine on which the replacement

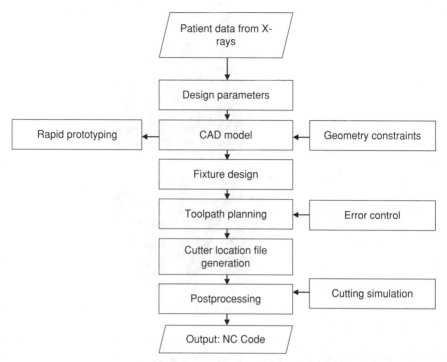

Figure 7.12 Flowchart for manufacture of femoral part of knee replacement. *Source:* Reproduced from Lee, J-N., Chen, H-S., Luo, C-W. and Chang, K-Y. (2009) Rapid prototyping and multi-axis NC machining for the femoral component of knee prothesis, *Life Sciences Journal* **6** (3), 73–77.

joint is to be machined. Preparation of this part program is usually done away from the machining area. Should the operator want to verify the part program, the machine tool controller can display the tool path on a graphic display screen. This assumes that the machine has a direct link to the central computer on which the part program was processed. Alternatively, the part program can be verified using other commercial software such as VERICUT. This part program is generated.

These procedures are general for most joint replacement manufacture. As an example, the femoral component of a knee joint replacement is discussed here. It had to replace the lower surface of the femur and the groove into which the patella fits. Its location and that of the other knee parts to be replaced are illustrated in Figure 7.13.

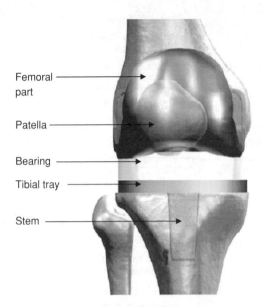

Femoral part

Patella

Bearing

Tibial tray

Stem

Figure 7.13   Illustration of total knee replacement parts.
Source: Reproduced from Lee, J-N., Chen, H-S., Luo, C-W. and Chang, K-Y. (2009) Rapid prototyping and multi-axis NC machining for the femoral component of knee prothesis, *Life Sciences Journal* **6** (3), 73–77.

The design features of the femoral component to be produced are shown in Figure 7.14. Figure 7.15 shows a simulation of the machining of the tibial-femoral surface. Another example of these CAD/CAM procedures has been described by Skalski and Grygoruk (2004) for the manufacture of a femur replacement. They began with a cylinder of diameter and length 89 × 260 mm of 316 type stainless steel. Using a CAD/CAM system, an endoprosthesis was produced from the cylinder. An intermediate part was made by electrodischarge machining (EDM). Then the initial 'rough' shape was obtained by milling with a flat cutter of 12 mm diameter.

A second stage of milling with a ball-ended cutter of 10 mm diameter yielded an implant with 0.2 mm allowance over its entire surface, except from the head cone which was left with an allowance of 0.5 mm. A third stage of milling with a ball-ended cutter of 8.0 mm diameter completed the manufacture. Milling was then repeated for the other faces of the endoprosthesis. A manually operated lathe was used to produce the head cone to its required surface roughness, and remove excess metal from the other sections of the part. Manual polishing completed the finishing.

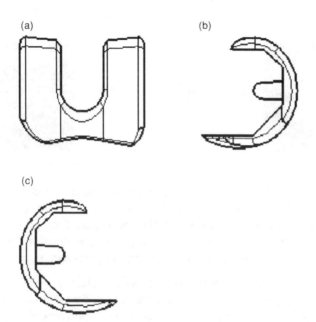

Figure 7.14   Design features of femoral part of knee replacement: (a) front, (b) left side sagittal, and (c) right side sagittal views.
Source: Reproduced from Lee, J-N., Chen, H-S., Luo, C-W. and Chang, K-Y. (2009) Rapid prototyping and multi-axis NC machining for the femoral component of knee prothesis, *Life Sciences Journal* **6** (3), 73–77.

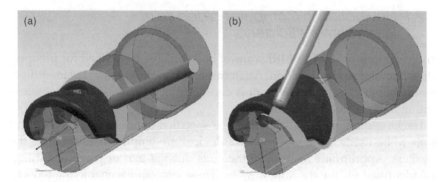

Figure 7.15   Finish machining of the (a) tibiofemoral surface and (b) fillet surface.
Source: Reproduced from Lee, J-N., Chen, H-S., Luo, C-W. and Chang, K-Y. (2009) Rapid prototyping and multi-axis NC machining for the femoral component of knee prothesis, *Life Sciences Journal* **6** (3), 73–77.

(a)   (b)   (c)

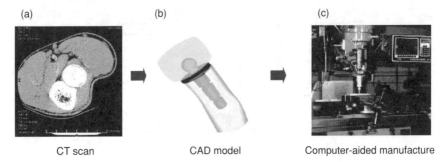

CT scan   CAD model   Computer-aided manufacture

Figure 7.16   Phases of manufacture of elbow joint replacement.
Source: Reproduced with permission from Skalski, K. and Grygoruk, R. (2004) Manufacture Process of Modular Radial Head Mobile Endoprosthesis and its Instruments. Warsaw University of Technology, Proceedings of the III International Conference on Advances in Production Engineering, Part III, pp. 41–46.

The manufacture of an elbow joint replacement has also been reported by Skalski and Grygoruk (2004). After having obtained a CT image of an elbow joint, they used a CAD/CAM system to produce a solid model containing the main anatomical features of the elbow; the CAM system was subsequently used to generate the tool paths to machine the replacement joint on a CNC machine. Figure 7.16 shows the phases used in the design and manufacture.

7.7 Navigation

7.7.1 Navigation in Computer-aided Joint Replacement Surgery

Three-dimensional CT or MRI scans of bone images are needed for preoperative planning and navigation. Anatomic and bone landmarks are identified. A suitable model for the implant is chosen from a database. Its orientation within the bone model is selected.

Tracking markers are attached to the specified bones and to the surgical tools to be used. Positional tracking devices are employed to monitor the markers. Appropriate features of the tools, such as axis or plane of cutting, are identified for the tracking devices. These procedures enable the tools to be positioned relative to the bones involved. These positions are presented both in numeric and visual form for subsequent navigation in surgery.

Jaramaz and DiGioia (2004) place these surgical procedures into (a) preoperative, and (b) intraoperative systems.

In the former, three-dimensional CT or MRI images are used to create three-dimensional models of bones and cross-sectional views. A reference system for the bones is defined. For example, in total hip replacement, the pelvis might be used in the definition of these locations and reference points. This preoperative procedure enables planning the positioning of the acetabular cup. The appropriate location of the implant, which is chosen from a database, is then defined. Further procedures include markers attached to specified bones, which are used to track their position during planning the surgery needed.

Trackers with anatomic reference methods are also used in intraoperative models. Images retrieved from fluoroscopy can be used. An alternative to image-reliant methods involves identification of prominent bone positions. In total hip replacement (THR), the centre of the hip could be defined as one such point for the centre of rotation of the femur relative to the pelvis. Other landmark points are supplied from tracking probes on the hip bone, which provide a bone reference system. The plan for surgery is then based on the bone reference framework.

Jaramaz and DiGioia (2004) have described in more detail navigation in THR. They emphasize the significance of anatomic reference locations and alignment of the implant. The anatomic references are needed to orientate the parts of the THR; for the acetabular cup the anterior pelvic plane is used as an anatomic reference system. The coordinates for the femur are given by the line passing from the centres of the femoral head and of the knee. The lesser trochanter and the medial and lateral posterior condyle points are used to specify the transverse plane. They define the centre of the knee as the mid-condyle point translated with the transverse plane, the latter plane being shifted in the perpendicular direction and passing through the femoral head centre. The orientation of the stem is derived from the two angles of the stem axis, and the 'version' angle of the stem is specified as the angle between the axis of the stem neck and the transverse plane of the femur. The term 'version' describes the extent to which an anatomical structure is rotated forwards ('anteversion') or backwards ('retroversion') relative to a specified reference point (Fabeck *et al.* 1999). The navigation system provides first simulation of the range of motion possible in preoperative planning. In this part the components of the hip joint replacement can be chosen. Then the movements of the replacement joint can be simulated. Any adjustments can be made to optimize its function in its various modes of flexion and extension, abduction and adduction, internal and external rotation, and rotation in flexion. The information so gathered is then stored for implementation in the navigation for THR surgery.

In the navigation system described by Jaramaz and DiGioia (2004) markers for tracking are attached to the pelvis. Bone (or any tool) position is tracked optically to less than one millimetre accuracy by use of means of light-emitting diodes (LED) and a camera. A probe is used to touch the surface of the bone in the region of the acetabulum in order to register key points on the pelvis. These data can then be compared with the CT scans gathered for the pre-operative planning phase. The planned orientation of the acetabular cup can then be checked. Similar navigation systems are employed to cover femoral tracking and stem placement.

The navigation provides an accurate means of checking the final position of orientation of the implant relative to earlier planning.

Stulberg *et al.* (2004) have discussed navigation systems for total knee replacements (TKR). Their implementation is aimed at greater accuracy. Reliance on preoperative planning can be reduced by image-free TKR navigation. This system requires that the centres of the hip, knee and ankle joints are first determined. The centre of the femoral head is defined by flexing, extending, abducting, adducting and rotation of the femur. These kinematic movements provide the data from which the surgeon collects information concerning motion of the hip. Next the ankle joint undergoes flexion and extension, from which information on its centre is derived. Movement of the ankle joint can be observed from a user interface. Flexion and extension of the knee provide the data needed to register the centre of the knee joint. Surface registration of the knee and ankle joints is also performed. This procedure enhances the accuracy of the calculations of the centres of the joints described above.

Reference points for the tibia and femur are registered. The size and rotation of the femoral part is determined. Further procedures are used to confirm the centre of the ankle joint. The navigation registration procedures for TKR are then complete. The tibial cutting block is positioned, secured by pins to the tibia. Its location is checked with the navigation systems. From the computer screen, the amount of tibial resection needed can be decided. Resectioning is then performed. Similar procedures are applied to resectioning of the femur.

7.7.2 Navigation in Robotic Surgery

Taylor *et al.* (2009) advocate the use of decision-support software in which the kinematics and stresses that affect replacement joints can be modelled computationally in order to provide more information in the planning of TKR surgery. Similarly Casino *et al.* (2009) have discussed new computational methods for navigation systems that can be used for evaluating the kinematics associated with OA-affected and reconstructed knees. Figure 7.17 shows an example of navigation system equipment used in joint replacements.

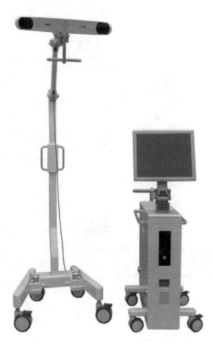

Figure 7.17 Navigation system for joint replacement.
Source: Reproduced with permission from Nakashima Medical Co.

Ball *et al.* (2009) also claim that computer-aided navigation enables consistent accurate placement of unicompartmental knee prostheses.

Knee-joint kinematics associated with flexion, extension and other gait movements such as walking, stair ascent and descent, and squatting have been assimilated into a data-logging system based on electrogoniometry developed by Rowe *et al.* (2009).

The accuracy for clinical purposes of hip navigation software-measurement systems for acetabular alignment has been confirmed by Arachchi *et al.* (2011).

7.8 Robotics

The use of robotics in joint replacements has been investigated initially to improve the quality of milling of bone. With close contact between bone and a cementless implant, bone growth is improved, as is stability for the replacement. An appreciation of the main elements of surgical robotics is a useful foundation on which to base this application for such prostheses. Hardware is illustrated in Figure 7.18.

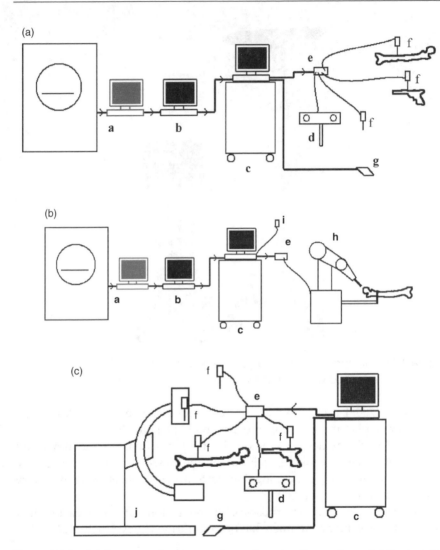

Figure 7.18 (a) Preoperative image-based navigation system. (b) Preoperative image-based robotic system. (c) Intraoperative image-based navigation. (d) Image free navigation.
Source: Reproduced from DiGioia, A.M., Jaramaz, B., Picard, F. and Nolte, L.P. (eds) (2004) "Computer and Robotic Assisted Knee and Hip Surgery" Oxford University Press, Oxford.

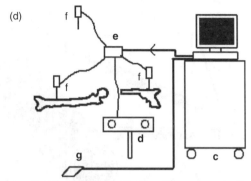

(d)

Notation: a: imaging station; b: planning station; c: navigation cart;
d: tracking device; e: controller; f: tracking marking; g: foot pedal;
h: robot; i: safety switch; g: fluoroscope.

Figure 7.18   (*Continued*)

In imaging for robotics, CT, MRI and fluoroscopy are most widely used. These techniques have been described in Chapter 4.

UNIX or Windows systems are widely used. The computer, monitor, keyboard mouse, power transformer and tracking controller can be mounted on a moveable cart. From the trackers, positional measurements of the appropriate anatomy and surgical tools are presented to the surgeon, by means of a monitor indicating the present, and possible future, positions. The surgeon's decision can be communicated to the computer by various procedures such as foot pedals, touch screens and voice commands. Tracking navigation systems such as infrared light-based methods have been discussed by Langlotz (2004).

The surgical instrument is moved by the robot 'manipulator'. This is made up of several joints, the movement of each of which is computer controlled. The joints are moved by drive mechanisms that consist of a motor and transmission elements such as a lead screw and connecting rods. Encoders give feedback on position for each joint.

Figure 7.19 illustrates these aspects of the robot, in which 'link 1' and 'link 2' indicate its structural elements.

The motion of link 1 relative to link 2 by revolute is shown in Figure 7.19. The term 'kinematic design' is used to describe the means by which the joints and links of the robot are connected in order to provide motion of the tool (that is, the surgical instrument) at some location away from its base. The

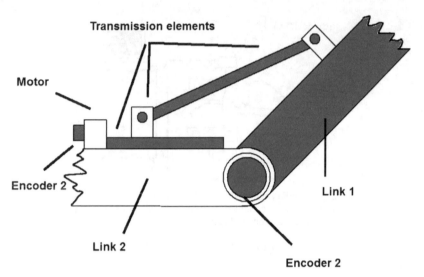

Figure 7.19   Illustration of robot manipulator for surgery.
Source: Reproduced from DiGioia, A.M., Jaramaz, B., Picard, F. and Nolte,
L.P. (2004) (eds) "Computer and Robotic Assisted Knee and Hip Surgery" Oxford
University Press, Oxford.

position of the tool, and its orientation, is determined by the computer that
controls the robot from the combination of information on the kinematic
design and positional feedback from the joint encoders. The computational
exercises associated with these procedures are known as 'forward kinematic
solutions' for the robot. The computer can also determine the 'inverse kinematic solution' to find the joint positions for a required tool position and
orientation. These orthopaedic surgical robots incorporate high gear reduction in their joint transmission elements. The robot then tends to be stiff,
precise and slow in speed and acceleration.

Typical "end effectors" are the surgical cutters, saws or drill guides that are
manipulated by the robot. The former term may also be used to describe the
device by which the tool is attached to the robot.

The various controls used in the robotic system can now be summarized,
see Figure 7.20. Paramount are safety and consistency.

Software controlling the position of the robot joints, and the kinematics of
the robot defines the position and orientation of the surgical tool. The joint
trajectories can be produced. An 'application programming interface' is often

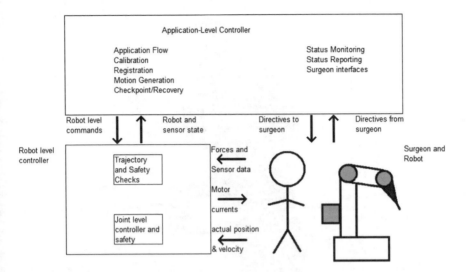

Figure 7.20   Controller for robotic surgery.
Source: Reproduced from DiGioia, A.M., Jaramaz, B., Picard, F. and Nolte, L.P. (2004) (eds) "Computer and Robotic Assisted Knee and Hip Surgery" Oxford University Press, Oxford.

incorporated within the robot controller software to provide the commands needed to the trajectory control software.

The power and interlocks to the entire robot system are controlled by electrical interface controls such as microprocessors. A useful description of robot technology may be found in Todd (1986).

Figure 7.21 shows an industrial robot for robotic surgery that incorporates these features.

7.8.1 Robotics-Assisted Total Knee Replacement (TKR)

A robotic system for TKR reported by Siebert *et al.* (2004) uses an industrial robot, CT images, and interactive computer-assisted preoperative planning. High accuracy in the positioning of the implant is claimed. The procedure involves placement of femoral and tibia pins in the form of self-tapping screws. The pins act as markers that can be detected by CT. The position enables the robot to be oriented spatially and appropriate geometric calculations to be made.

The CT data are transferred to a computer-based planning station. Anatomical prominences and axes of the femur and tibia can be noted

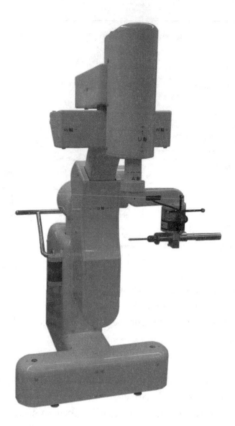

Figure 7.21   Example of industrial robot used in joint replacement.
Source: Reproduced by permission of Nakashima Medical Co.

and their mechanical positions can be calculated in the frontal and sagittal planes. The size and position of the implant can be chosen. The regions for bone milling are identified. Data are transferred to the robot control unit before surgery. The knee joint is screwed to a frame, to which are fitted light-emitting diodes, which monitor undesired micro-movements. Robotic milling is then controlled by the surgeon to produce accurately shaped and smooth bone surfaces.

In contrast with these claims, Cobb *et al.* (2009) suggest that computer-assistance in knee replacements (KR) surgery remains contentious. They undertook a five-year trial of a robotic system for unicompartmental knee replacement. The aim was improvement in accuracy for technical tasks associated with the surgery, such as angular alignment of the replacement prosthesis and associated tibio-femoral alignment in the coronal plane. Little significant benefit in the robotic route compared with conventional methods was observed. Others have found improvements in function and flexion associated with guided knee motion arthroplasty for patients aged about 70 with a BMI of 30.4 (Arbuthnot and Brink 2009).

Haptic robotics that replace conventional manual methods for unicompartmental knee replacement have been described by Jinnah *et al.* (2009). The robotic arm followed precisely a preoperative plan prepared from CT scans, enabling accurate bone resection. A robotically controlled arm, which enabled accurate, bone-sparing preparation of the femur and tibia was also implemented by Roche *et al.* (2009) in minimally invasive unicompartmental knee arthroplasty.

As an example, the industrial procedures used in computer-aided knee joint replacement are presented in Figures 7.22 (a) to (d). Figure 7.22 (a) shows segmentation of the bone into discrete parts.

The mechanical axis is defined in Figure 7.22 (b). The positioning of the implant is presented in Figure 7.22 (c). Finally, Figure 7.22 (d) indicates how the alignment is checked.

7.8.2 Robotics-Assisted Total Hip Replacement (THR)

Cementless THR can be assisted through robotics. In the methods described by Bauer (2004), CT scans can be used to fix the positions of titanium locator pins for example on the medial or lateral condyles (two pins), or on both femoral condyles and greater trochanter (three pins).

During the computer-aided planning stage a three-dimensional model of the proximal femur and joint can also be produced and viewed on a monitor. From these images and information, the appropriate size and position of the implant can be chosen, including the correct size of head and neck length. An outline of the recommended cutting (reaming) path is provided. A probe is guided to the centres of the hip pins. The distances and angles of the pins are calculated by the robot, and compared to the data from the CT scan slices. When the data all match, the probe is removed and a high-speed cutting tool is attached to a motor. After the cutter has been guided to the bone front, robotic cutting begins. On completion of reaming, the robot is taken from the operating table, the proximal fixator is removed. The implant is placed into the bone.

Pinless robotic THR uses the proximal femur and femoral neck as the registration surfaces. The leg positioning and fixation are similar to that of the pin-based method.

(a)

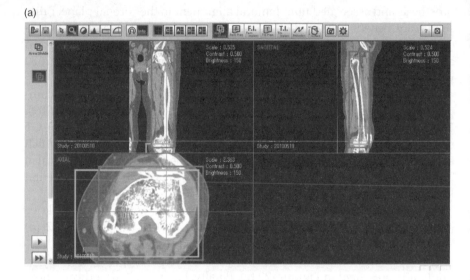

(b)

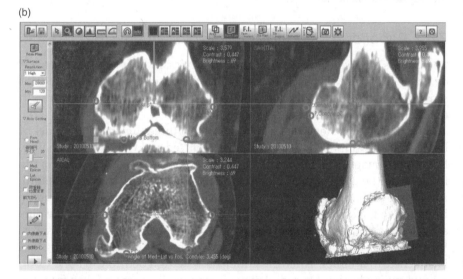

Figure 7.22 (a) Segmentation. (b) Definition of the mechanical axis. (c) Positioning of the Implant. (d) Check on alignment.
Source: Reproduced by permission of Nakashima Medical Co.

(c)

(d)

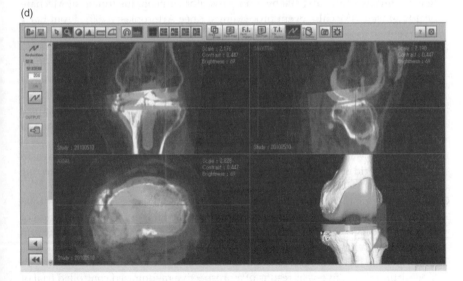

Figure 7.22 (*Continued*)

A monitor for the bone movement is attached. A computer enables the surgeon to position a digitizer arm attached to the robot to locations on the femoral neck. Reaming of the femoral canal then proceeds. Finally, the acetabular component and the femoral stem of the THR are placed in position.

References

Agarwal, Y., Frehill, B., Crocombe, A.D. *et al.* (2009) Stability of tibial implants in augmented knee arthroplasty. Institution of Mechanical Engineers (IMechE), event proceedings, Knee Arthroplasty 2009: From Early Intervention to Revision, pp. 285–288.

Arachchi, S., Augustine, A., Deakin, A. and Rowe, P. (2011) Technical validation of the accuracy of the OrthoPilot cup navigation software-measurement of native acetabular alignment. Institution of Mechanical Engineers (IMechE), event proceedings, Engineers and Surgeons: Joined at the Hip III, pp. 75–78.

Arbuthnot, J.E. and Brink, R.B. (2009) Early experiences, successes and pitfalls with a new guided knee motion arthroplasty intended to improve function and flexion – a review of the first 100 by a non-inventor surgeon. Institution of Mechanical Engineers (IMechE), event proceedings, Knee Arthroplasty 2009: From Early Intervention to Revision, pp. 211–213.

Ball, S., Windley, J. and Nathwani, D. (2009) The importance of alignment in knee replacement surgery – can we deliver surgical precision? Institution of Mechanical Engineers (IMechE), event proceedings, Knee Arthroplasty 2009: From Early Intervention to Revision, pp. 295–297.

Bauer, A. (2004) Total hip replacement – robot assisted technique, in *Computer and Robotic Assisted Knee and Hip Surgery* (eds A.M. DiGioia, B. Jaramaz, F. Picard and L.P. Nolte), Oxford University Press, Oxford, pp. 83–95.

Bougherara, H., Mahboob, Z., Miric, M. and Youssef, M. (2009) Finite element investigation of hybrid and conventional knee implants. *International Journal of Engineering* 3 (3), 257–266.

Casino, D., Lopomo, N., Martelli, S. *et al.* (2009) Intra-operative assessment of total knee replacement: a kinematic study adopting a navigation system. Institution of Mechanical Engineers (IMechE), event proceedings, Knee Arthroplasty 2009: From Early Intervention to Revision, pp. 271–274.

Cobb, J., Henkel, J., Rodriguez, F. and Davies, B. (2009) Accuracy improves outcome in knee arthroplasty: five-year results of a prospective randomised controlled trial of robotic unicompartmental knee replacement. Institution of Mechanical Engineers (IMechE), event proceedings, Knee Arthroplasty 2009: From Early Intervention to Revision, pp. 211–213.

Dandachli, W., Najefi, A., Lenihan, J. et al. (2011) Quantifying the contribution of pincer impingement to femoro-acetabular impingement. Institution of Mechanical

Engineers (IMechE), event proceedings, Engineers and Surgeons: Joined at the Hip III, pp. 21–23.

Delaunay, B. (1934) Sur la sphère vide. *Bulletin of Academy of Sciences of the USSR* **6**, 793–800.

DiGioia, A.M., Jaramaz, B., Picard, F., and Nolte, L.P. (eds.) (2004). "Computer and Robotic Assisted Knee and Hip Surgery", Oxford University Press, Oxford, U.K.

Eosoly, S., Brabazon, D., Lohfeld, S. and Looney, L. (2009) Selective laser sintering of hydroxyapatite/poly-ε-caprolactone scaffolds *Acta Biomaterialia* **6** (7), 2511–2517.

Fabeck, L., Farrokh, D., Tolley, M. *et al.* (1999) A method to measure acetabular cup anteversion after total hip replacement. *Acta Orthopaedica Belgica* **65** (4), 485–491.

Fisher, J., Al Hajjar, M., Stewart, T. *et al.* (2011) Pre-clinical evaluation of the tribology of hip prostheses under adverse conditions. Institution of Mechanical Engineers (IMechE), event proceedings, Engineers and Surgeons: Joined at the Hip III, pp. 51–54.

Greenwald, S., Morra, E. and Rosca, M. (2009) Kinematics performance of high flexion knee designs. Institution of Mechanical Engineers (IMechE), event proceedings, Knee Arthroplasty 2009: From Early Intervention to Revision, pp. 37–38.

Hellwig, F., Tong, J. and Hussell, J.G. (2011) Femoroacetabular impingement due to acetabular and femoral anteversion: a finite element study. Institution of Mechanical Engineers (IMechE), event proceedings, Engineers and Surgeons: Joined at the Hip III, pp. 11–18.

Horáček, M., Charvát, O., Pavelka, T. *et al.* (2010) Medical implants by using RP and investment casting technologies. Proceedings of the 69th World Foundry Congress, Hangzhou, China, pp. 107–111.

Jaramaz, B. (2004) Images devices, computers, peripherals, interfaces, in *Computer and Robotic Assisted Knee and Hip Surgery* (eds A.M. DiGioia, B. Jaramaz, F. Picard and L.P. Nolte), Oxford University Press, Oxford, pp. 49–51.

Jaramaz, B. and DiGioia, A.M. (2004) Total hip replacement – navigation technique, in *Computer and Robotic Assisted Knee and Hip Surgery* (eds A.M. DiGioia, B. Jaramaz, F. Picard and L.P. Nolte), Oxford University Press, Oxford, pp. 79–82.

Jaramillo-Botero, A., Blanco, M., Li, Y., McGuinness, G., and Goddard, W.A. (2010) First-Principles Based Approaches to Nano-Mechanical and Biomimetic Characterization of Polymer-Based Hydrogel Networks for Cartilage Scaffold-Supported Therapies. Journal of Computational and Theoretical Nanoscience, 7 (7). pp. 1238–1256. ISSN 1546-1955

Jinnah, R.H., Lippincott, C.J., Horowitz, S. and Conditt, M.A. (2009) The learning curve of robotic-assisted UKA. Institution of Mechanical Engineers (IMechE), event proceedings, Knee Arthroplasty 2009: From Early Intervention to Revision, pp. 263–265.

Langlotz, F. (2004) Localizers and trackers for computer assisted freehand navigation, in *Computer and Robotic Assisted Knee and Hip Surgery* (eds A.M. DiGioia, B. Jaramaz,

F. Picard and L.P. Nolte), Oxford University Press, Oxford University Press, Oxford, pp. 51–53.

Lee, J.-N., Chen, H-S., Luo, C-W. and Chang, K-Y. (2009) Rapid prototyping and multi-axis NC machining for the femoral component of knee prothesis. *Life Sciences Journal* **6** (3), 73–77.

McGeough, J. A., McLean, A. J. and Keating, J. (2003) Mechanical properties of human knee meniscus assisted by sutures or meniscal arrows. *Acta of Bioengineering and Biomechanics* **4**, 335–336.

McMahon, C. and Browne, J. (1995) *CADCAM: From Principles to Practice*, Addison-Wesley. Reading, MA, U.S.A.

Mortenson, M.E. (2006) *Geometric Modelling*, 3rd edn, Industrial Press. New York.

Picard, F., Moody, J.E., and DiGioia, A.M. (2004). Chapter 1 – A history of computer-assisted orthopeadic surgery of the hip and knee, in *Computer and Robotic Assisted Knee and Hip Surgery DiGioia*, (eds A.M. DiGioia, B. Jaramaz, F. Picard, and L.P. Nolte), Oxford University Press, Oxford, U.K.

Phillips, A.T.M., Howie, C.R., and Pankaj, P. (2008). Biomechanical evaluation of anterolateral and posterolateral approaches to hip joint arthroplasty. *Journal of Bone and Joint Surgery (British) Proceedings*, 90-B, 547–548

Raja, V. and Fernandes, K.J. (2008) *Reverse Engineering – An Industrial Perspective*, Springer. London.

Roche, M.W., Augustin, D. and Conditt, M.A. (2009) Robotically guided UKA: outcomes of initial series. Institution of Mechanical Engineers (IMechE), event proceedings, Knee Arthroplasty 2009: From Early Intervention to Revision, pp. 267–269.

Rowe, P.J., Smith, J. and Padmanaabhan, V. (2009) Towards phase III: multi-centred RCTS of functional outcome of TKA using flexible electro-goniometry: the challenge for the next decade. Institution of Mechanical Engineers (IMechE), event proceedings, Knee Arthroplasty 2009: From Early Intervention to Revision, pp. 71–74.

Sandholm, A., Schwartz, C., Pronost, N. *et al.* (2011) Evaluation of a geometry-based knee joint compared to a planar knee joint. *International Journal of Computer Graphics* **27** (2), 161–171.

Siebert, W., Mai, S., Kober, R. and Heekt, P.F. (2004) Total knee replacement: robotic assistive technique, in *Computer and Robotic Assisted Knee and Hip Surgery* (eds A.M. DiGioia, B. Jaramaz, F. Picard and L.P. Nolte), Oxford University Press, Oxford, pp. 127–138.

Simon, F.D., Hoeltzel, D.A. and Gustilo, R.B. (1985) A computer-aided method for describing human knee geometry. Proceedings on the 1985 Symposium of Biomechanics, pp. 105–108.

Skalski, K. and Grygoruk, R. (2004) Manufacture process of modular radial head mobile endoprosthesis and its instruments. Warsaw University of Technology, Proceedings of the III International Conference on Advances in Production Engineering, Part III, pp. 41–46.

Smith, J.O., Sengers, B.G., Aarvold, A. *et al.* (2011) Tissue engineering strategies to extend the orthopaedic application of tantalum trabecular metal. Institution of Mechanical Engineers (IMechE), event proceedings, Engineers and Surgeons: Joined at the Hip III, pp. 89–92.

Stulberg, S.D., Saragaglia, D. and Miehlke, R. (2004) Total knee replacement: navigation technique intra-operative model system, in *Computer and Robotic Assisted Knee and Hip Surgery* (eds A.M. DiGioia, B. Jaramaz, F. Picard and L.P. Nolte), Oxford University Press, Oxford, pp. 127–138.

Taylor, M., Zachow, S., Heller, M. *et al.* (2009) Decision support software for orthopaedic surgery. Institution of Mechanical Engineers (IMechE), event proceedings, Knee Arthroplasty 2009: From Early Intervention to Revision, pp. 215–219.

Tlusty, G. (1999) *Manufacturing Processes and Equipment,* Prentice Hall, Upper Saddle River NJ.

Todd, D.J. (1986) *Fundamentals of Robot Technology*, Kogan Page, London.

Wang, J.J. and Wu, Q.Q. (2009) Numerical prediction on mechanical contacts: Vangard knee joint replacements tested in the displacement-controlled ProSIM Simulator. Institution of Mechanical Engineers (IMechE), event proceedings, Knee Arthroplasty 2009: From Early Intervention to Revision, pp. 289–292.

8

Joint Replacement

8.1 Introduction

The clinical introduction and use of any orthopaedic implant depend on achieving approval through two separate but related routes. Final clinical approval is obtained in the United States from the Food and Drug Administration (FDA) and in the European Union through the CE approval system, both currently receiving considerable criticism (Cohen 2012). Approval requires evidence of clinical studies and relevant laboratory studies, which will include evidence of manufacturing and quality processes. These systems have been shown to be deficient, with many implants being introduced with little clinical evidence. As a result, there is considerable concern, for instance, about the introduction of modern metal-on-metal bearings (Wilkinson 2012) and as discussed more fully below (Bolland *et al.* 2011). The Medicines and Healthcare products Regulatory Agency (MHRA) (2010) also issued a 'medical device alert' on this type of implant.

In contrast, the materials and manufacturing processes used in joint replacements are subject to rigorous control over quality and performance. Many of these requirements fall under the aegis of ISO (International Organization for Standardization). Regulatory bodies in most countries impose their own conditions for manufacturing. For example, in the United Kingdom, the British Standards Institution (BSI) specifies many conditions. In the United States, the American Society for Testing and Materials (ASTM) documents its requirements. The European Commission regularly issues directives. Medical associations set out the training required and codes to be followed in joint replacement surgery, such as ISO 7206.

The Engineering of Human Joint Replacements, First Edition. J.A. McGeough.
© 2013 John Wiley & Sons, Ltd. Published 2013 by John Wiley & Sons, Ltd.

The list of standards and legal requirements that may be found in ASTM, ISO, and BSI reports is extensive. Some examples of the specifications for the joints investigated in earlier chapters are described below.

The following ISO documents describe the characteristics and test methods for metallic materials:

- ISO 5832-1:2007 part 1: wrought stainless steel;
- ISO 5832-3:1996 part 3: wrought Ti6Al4V;
- ISO 5832-4:1996 part 4: cobalt-chromium-molybdenum casting alloy.

Ultra-high molecular weight polyethylene (UHMWPE) in moulded form (for example sheets and rods) is covered in ISO 5834-2:2001. The tests methods for morphology of UHMWPE are defined in ISO 5834-5:2005 part 5. Ceramic materials are documented in ISO 6474. Part 1:2010 deals with ceramics bone substitute materials based on high purity alumina as a replacement for bone and parts of joint prostheses. Its part 2-2:2012 reports on composite materials based on high purity alumina matrix with a zirconia reinforcement. Acrylic cements are specified in ISO 5833:2002. The requirements for curing resin cements are set out. The respective amounts of sterile powder and liquid to be mixed at implantation are given. This ISO deals with two types of cement, for use with a syringe, or as a dough for fixation of the implant. In addition, the physical and mechanical conditions, and the packaging of the cement in its sterile condition, are specified.

Other requirements in manufacturing are also found in the ISO documentation. For instance, those for plasma-sprayed coatings (for unalloyed titanium) are available in ISO/DIS 13169-1. Outwith manufacturing, clinical procedures for receiving, storage, transport, handling, cleaning, and sterilization of the implants are given in ISO 8828:1988.

The ASTM F 1378-12 specification covers shoulder prostheses employing glenoid and humeral components. It discusses devices that are fully constrained, partially constrained and unconstrained. Customized applications that are tailored to the needs of a specific patient are excluded. Conditions are given to deal with the mechanical strength, corrosion resistance, biocompatibility, wear, and range of motion of the shoulder-joint replacement. ASTM F 1829-98 (2009) sets out test methods on the effects of materials, manufacturing, and design variables for metal-backed glenoid prostheses, and locking mechanisms for resisting static shear loading. ASTM WK28231 deals with elbow joint replacement.

Total wrist implants that provide articulation through radial and carpal components are described in ASTM F 1357-09. The biocompatible

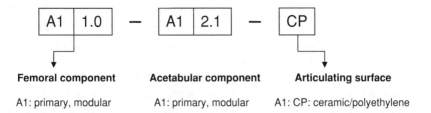

Figure 8.1 Example of hip joint prosthesis marking code.
Source: Based on BS EN 980:2008 (2008).

materials from which the implants should be manufactured are listed, along with their specific dimensions. They include unalloyed titanium, its alloys (Ti6Al4V), various cobalt-chromium-molybdenum alloys, and also solid ceramic implants. Tests for polymeric creep, wear, and range of motion are designated. Implants with ceramic-coated or porous-coated surfaces and one-piece elastomeric implants are excluded.

Finger joint implants are reported in ASTM F 1781-03 (2009). The standard deals with biocompatibility and test methods for elastomeric flexible hinge finger total joint replacements, made from one material in a single step moulding process.

ASTM F 2033-05 is the standard specification for total hip-joint prostheses and hip endoprostheses bearing surfaces made of metallic, ceramic, and polymeric materials.

ISO 7206-1 divides the prosthesis into a femoral and acetabular component, and defines conditions for the articular surfaces. Figure 8.1 illustrates some of its specifications. Procedures concerned with cement injection are covered under ISO 12891.

Design conditions which must be met include:

- Biocompatibility of materials and chemicals.
- Endurance, wear and degradation of the materials. ISO 7206-6:1992 describes methods for testing the endurance properties of the head and neck regions of stemmed femoral components, for both partial and total hip joint prostheses. The wear of total hip joint replacements is covered in ISO 14242.
- Effects of shape and dimensions of the implant within the body. Its metallic components have to satisfy ISO 7206-2 with requirements for surface finish (e.g. R_a not more than $0.05\,\mu m$ for articular surfaces), tolerance (diameters have to have a negative tolerance between 0.2 and 0.0 mm),

and sphericity (radial separation of not more than 10.0 μm). The polymeric parts are defined by ISO 7206-2 and ISO 7206-1. Surface finish has to be not more than 2 μm R_a for the acetabular cup articulating surfaces. The diameters should have a positive tolerance between 0.1 and 0.3 mm at 20 °C ± 2 °C. In sphericity, the radial separation should be not more than 100 μm. Similarly, the ceramic components have to be produced to their ISO requirements. Prior to use, the hip prostheses have to be tested under ISO 7206-4 and ISO 7206-8. The device should not fracture during 5×10^6 cycles of loading. The loading has to be cycled through a load of 300 N up to 2.3 kN at a frequency of 5 Hz in order to simulate the stress levels expected for a patient using the prosthesis for more than five years. For a minimum batch of eight femoral components, 75% have to pass these tests for acceptance to be granted:

- Effects and consequences of radiation and the electromagnetic spectrum on the implant.
- Influence of particulates and bacterial levels on function of the implant.
- Ease of removal and disposal of the implant.
- Analysis and procedures for preclinical trials, and subsequent postmarket surveillance (under a separate ISO 14971).
- Manufacturers supplying implants in a sterile state are expected to comply with existing assurance levels. They have to present this documentation, labelling their goods with the relevant ISO, and marking the items 'sterile'. When nonsterile implants are supplied, the manufacturer is expected to specify the most appropriate form of sterilization.

ASTM F 1223 deals with total knee replacements, offering a database of product functionality to assist with surgical choice. It draws attention to the variation in total knee replacement product design, and the range of constraints depending on geometric and kinematic interactions among the components. Methodologies and limitations are described, including preoperative and postoperative capability, activity and lifestyle of the patient. ISO 7207-1:2007 classifies the femoral, tibial, and patellar parts of knee joint replacements, involving the bearing surfaces, with definition and dimensions for components. Wear of total knee joint prostheses is discussed in ISO 14243.

Total ankle replacements are specified in ASTM F 2665-09. The standard covers articulation that allows talar and tibial components a minimum of 15° dorsiflexion and 15° to 25° of plantar flexion. The specification includes ankle components for primary and revision surgery, with modular and nonmodular designs. Bearing components are discussed for use with and without cement, as are materials and prosthesis geometry.

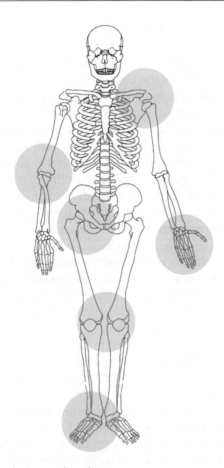

Figure 8.2 Joint replacement regions.
Source: Reproduced with permission from Nakashima Medical Co.

For these standards to be attained, much development of joint replacements has taken place over many years. Significant advances for each of the joints, indicated in Figure 8.2, are described below. This progress has led to the availability of the devices from industrial manufacturers worldwide, examples of which are also presented.

8.2 Shoulder

The main causes of total shoulder replacement (TSR) (also known as total shoulder arthroplasty, or TSA) are fracture, OA, RA, post-traumatic arthritis, the combined effects of a tear of the rotator cuff tendon with arthritis and osteonecrosis.

Total shoulder replacement is less common than that of other joints such as hip and knee. For example, in the United States about 53 000 TSRs were performed in 2007, compared to more than 900 000 hip and knee arthroplasties, (according to the American Academy of Orthopaedic Surgeons 2011).

In 1893, a platinum and rubber replacement for the glenohumeral joint of the shoulder was attempted. In 1921, the fulcrum of the glenohumeral joint was replaced by the proximal fibula of a patient to account for loss of the proximal humerus. A further attempt at an arthroplasty was undertaken in 1933 in which parts of the humeral head and proximal shaft were resected. The proximal humerus was rounded off, and the components of the musculotendinous cuff were reattached.

Another attempt at shoulder replacement came in 1951, when the humeral head was replaced with an unconstrained Vitallium replacement. Successive improvements in 1974 involved glenoid resurfacing with a polyethylene implant. Total shoulder replacement with a constrained, so called reverse articulation implant was available in the 1970s, with limited success. Modular humeral components were produced in the 1980s to account for variations in the humeral anatomy, head diameter thickness and offset. Glenoid parts were designed for cementless fixation by use of screws and porous coatings applied to a metal backing for the polyethylene bearing surface.

Further advances in the 1990s were concerned with restoration of normal kinematic movement through the anatomical location and orientation of the humeral glenoid joint surfaces. Canale and Beaty (2007) observed that non-constrained prostheses should closely replicate normal bony anatomy.

The humeral component can be inserted without cement by initial press fitting and relying on a surface coating to ensure long term ingrowth and fixation, if the residual bone is of sufficient substance. Soft bone may require the implant to be secured with cement.

A shoulder replacement usually comprises a ball and socket joint. One part is fixed to the glenoid region of the scapula by screws (see Figure 8.3). The socket (glenoid) may be constructed from UHMWPE which can be backed by porous trabecular titanium alloy. The ball component, the humeral head, can be made from cobalt-chromium alloy. The stem component, the humerus, can be manufactured from metal alloys such as stainless steel, or alloys that include tantalum, titanium, vanadium, and cobalt-chromium. More recent designs have included surface replacements and ultra-short stems into the head, when the bone stock is adequate.

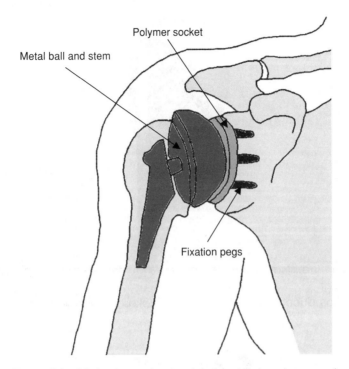

Polymer socket

Metal ball and stem

Fixation pegs

Figure 8.3 Main elements of a shoulder joint replacement.

To compensate for the absence of muscles to control shoulder movements there has been a resurgence in the concept of the reverse constrained shoulder where the glenoid becomes a ball and the humerus is made into a socket with specially designed implants.

Forces applied to the joint through the humerus are transmitted to the glenoid anchorage as bending moments in the stem that joins the ball or socket to the anchorage. The corresponding bending stress will be accompanied by shear stress. If the joint is fully constrained, these latter shear stresses may also be accompanied by either compressive or tensile stress. The purpose of the stem is to diminish prosthetic movement within the bone.

The forces that can be transmitted through the humerus can usually be calculated for normal practical conditions. Accidental conditions, for example if the elbow is suddenly struck or is leant on, are more difficult to analyse. In the design of these prostheses, high-strength screws in the glenoid region and stem are preferred as higher loads at the joint could loosen it from the scapula.

Figure 8.4 Industrial example of a shoulder hemi-arthroplasty joint replacement.
Source: Reproduced with permission from Nakashima Medical Co.

Figure 8.5 Industrial example of a total shoulder joint replacement.
Source: Reproduced with permission from Smith & Nephew, Inc.

An implanted biological scaffold adapted for shoulder-joint replacement, has been reported in tests on rabbits in whom the humeral head was removed (Inui *et al.* (2011)). The scaffold was treated in such a way as to develop the rabbit's own cartilage and bone stem cells. The researchers describe the formation of cartilage in the scaffolds within four months, thereby providing a fresh cartilage surface for the humeral head.

Revision of TSR arises from loosening, infection and wear of components of the prosthesis. Rotator cuff deficiency is often the cause, especially with patients with RA, and postoperative displacement of the tuberosities, which leads to eccentric loading on the glenoid component. The result is failure at the bone cement interface. The outcome from the shoulder joint prosthesis revision is difficult to predict, with rates of revision as low as 18%. Industrial examples of shoulder joint prostheses are shown in Figures 8.4 and 8.5.

8.3 Elbow

Wallace (1998) has reviewed the development of total elbow replacement (TER). In 1975 Ewald reported an unconstrained elbow arthroplasty. Its features included long humeral stem fixation and a polyethylene ulnar component. This device was found to be effective for RA cases. A similar, semiconstrained arthroplasty was developed by Souter (1977) and co-workers. Its characteristics were a short- (and later longer) stemmed humeral component that provided secure rotational fixation to the bone. It included a stirrup-shaped humeral prosthesis that offered enhanced anchorage of the component by using the contours of the medullary cavities of the medial and lateral supracondylar ridges, and an ulnar component that was first made in polyethylene. The device is described by Souter *et al.* (1985).

In 1972 Kudo began designing a series of TER. An example is his 'Type 4', which features a titanium alloy stem with a porous coating to promote biological fixation (1990).

A semi-constrained TER was designed in 1969 by Coonrad and Morrey. This TER duly underwent various modifications, such as introduction of a 7° hinge laxity or toggle in order to reduce bone-cement interference forces (1978). Later porous coatings were added to the distal humerus and proximal ulnar stems to enhance biological fixation, as well as bone cementing. Its features are depicted in Figure 8.6.

Stability of the normal elbow joint is usually achieved by its bony structure, neighbouring soft tissue, and the collateral ligaments. Deficiencies in

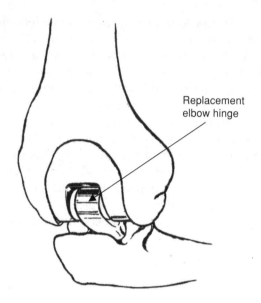

Figure 8.6 Total elbow replacement.
Source: Reproduced from Coonrad, R.W., and Morrey, B.F. (1998) Coonrad/Morrey total elbow: surgical technique. Zimmer Inc. product information, in Wallace, W.A. (ed,) *Joint Replacement in the Shoulder and Elbow,* Butterworth Heinemann, Oxford.

the soft tissue envelope can lead to various types of instability. Total elbow replacement loosening usually arises over a long time, often between four to six years. It is associated with bone erosion and resorption, followed eventually by bone fracture. Revision due to loosening is difficult due to the bone loss and subsequent difficulties of fixation. Usually a fully constrained implant is required to substitute for absent ligaments. Early assessment and treatment are the preferred route.

Failure of the TER arises mainly from defective fixation and from wear of the ultra-high molecular weight polyethylene (UHMWPE) components. With 'less-constrained' implants, fewer occurrences of failures have been reported (Wallace 1998).

Previously, constrained implants such as elbow hinges were prone to failure owing to deficiencies in the design or mechanisms of the hinge. Infection of the TER may arise if primary wound healing fails. Its onset is common in RA patients, the main recipients of TER, owing to their reduced immunity. Osteoporosis is often associated with RA; these inherent problems of bone

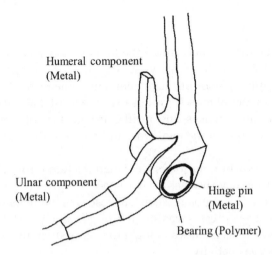

Humeral component
(Metal)

Ulnar component
(Metal)

Hinge pin
(Metal)

Bearing (Polymer)

Figure 8.7 Schematic illustration of main features of a hinged elbow joint replacement.

metabolism can also contribute to TER failure by resulting in poor ingrowth and fixation.

A schematic example of the main features of a hinged elbow joint replacement is shown in Figure 8.7.

Figure 8.8 shows another example of a hinged elbow joint replacement.

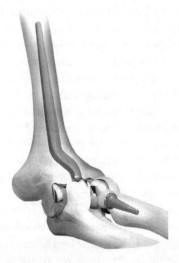

Figure 8.8 Industrial example of an elbow joint replacement.
Source: Reproduced with permission from Nakashima Medical Co.

8.4 Wrist

Total wrist joint replacement is aimed at retaining motion and offering a fixed fulcrum with stable fixation. The wrist is a complex arrangement of many bones with multiple planes of movement simultaneously at several levels. Moreover the number of muscle groups over the wrist that must be balanced to prevent dislocation makes this a difficult joint to replicate and replace. Stiffening (arthrodesis) of the joint was, and remains, the mainstay of wrist surgery.

The first replacement wrist joints were manufactured from silicone plastic (Swanson 1968) and were used as flexible spacers to prevent the joint surfaces from rubbing and causing pain. Their purpose was the relief of pain due to arthritis, which in severe cases was found to be effectively destroying the wrist joint. Wrist-joint replacement was sought to restore strength and movements needed for everyday activity.

Wright (2007) has discussed implants that are available, including the silicone spacer prostheses designed by Swanson. They do not require fixation by poly(methyl methacrylate) (PMMA), and involve little bone resection. Nonetheless, 10 to 52% of these prostheses have been known to fracture. This 'Swanson' implant has been recommended for those who do not need great use of their hands, or who have insufficient bone stock. Otherwise wrist joint replacements based on 'metal on plastic' designs may be employed.

Wright (2007) proceeds to review other wrist replacement designs including constrained devices by both Meuli (1984) and Volz (2007) which enable large forces to be transmitted to the implant. He has also described a semi-constrained biaxial porous-coated wrist replacement. With this device, cement-fixation can be improved or cementing made unnecessary; however, with both these types of implants, loosening has occurred. Figure 8.9 illustrates the main parts of a wrist-joint replacement.

Figure 8.10 shows the main features of a wrist-joint replacement. The end of the radius bone of the forearm (radial component) is fitted with a flat metal piece placed in front of the radius. Its stem penetrates into the canal of the bone. To this metallic radial component is then attached a plastic cup to form a socket for the replacement wrist joint. A second 'globe-shaped' metallic part fits into the plastic end of the radius, the 'distal component', and is used to replace the small wrist bones. It is attached by two metal stems that fit into the hollow bone marrow cavities of the carpal and metacarpal bones of the hand. This ball and socket replacement enables wrist movement in all directions. Figures 8.11 and 8.12 show further features of the wrist joint replacement, including an X-ray of its location.

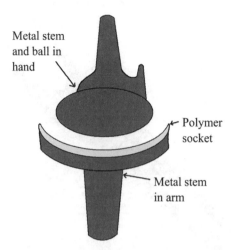

Metal stem
and ball in
hand

Polymer
socket

Metal stem
in arm

Figure 8.9 Cross-section of components of a wrist joint replacement.

Figure 8.10 Industrial example of wrist joint replacement.
Source: Reproduced from DePuy Orthopaedics, Inc. (www.depuy.com).

Figure 8.11 Industrial example of a ball-and-socket wrist replacement. *Source:* Reproduced by permission from Zimmer Inc.

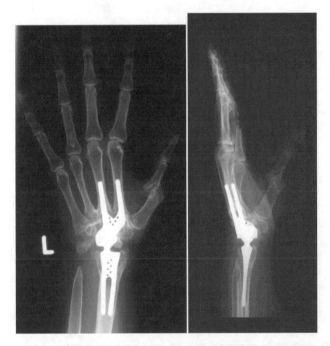

Figure 8.12 Postoperative X-ray of an implanted wrist joint replacement. *Source:* Reproduced by permission from Zimmer Inc.

8.5 Fingers

Fingers, thumb and wrist can all be affected by rheumatoid arthritis (RA), which destroys joint cartilages and erodes joints causing extensive deformities. The metacarpophalangeal are affected first, and subsequently the distal, joints. Osteoarthritis (OA) wears away the lining of the joints and affects the base of the thumb and the distal interphalangeal joints. It can cause the formation of spurs, fragmentation of the cartilages and limitation in motion. Inflammation and considerable pain are the consequences. Movement of hands and fingers becomes limited. Other adverse painful effects can arise from psoriatic arthritis, gout and synovitis (inflammation of the synovial membrane that lines the joints). The deformities at the fingers can even be caused by the normal forces that arise with everyday use of the finger joints.

The extrinsic flexors and extensors still act on the joints damaged by RA. The intrinsic muscles become tight as the lateral bands of the extensor hood become displaced. The long extensor or long flexor tendons become ruptured, creating a complex imbalance and characteristic deformities, which can be predicted for the deforming forces. Wright (2007) also reports that proximal interphalangeal arthroplasty of the central two finger digits can be achieved, as the digits on either side can provide lateral stability. However he also reports failure of cemented proximal interphalangeal arthroplasty devices, just over two years after surgery. He concludes that, at present, cemented articulated arthroplasty devices do not provide sufficient lateral stability in the radial proximal interphalangeal joints. The interphalangeal joints rely heavily on the soft-tissue structures gliding smoothly and predictably over the implant, although this has not yet been achieved.

Two anatomical components are usually used to replace the finger joint. The ends of the bone are removed. The stem of the replacement, usually made from titanium alloy, is inserted into the intramedullary marrow of each bone. The other part of the stem is made of UHMWPE plastic. Subsequent to finger joint replacement (FJR) the stability, looseness and range of movement of the implant are monitored.

Swanson (1968) describes his development of proximal interphalangeal joint (PIP) arthroplasty for patients suffering from RA and trauma. Biologically inert silicone elastomers were used as a main joint replacement material and the design is based on essentially a soft tissue spacer rather than a machined joint. He proceeds to describe replacement of the carpometacarpal joint of the thumb, a joint commonly affected by osteoarthritis, with an

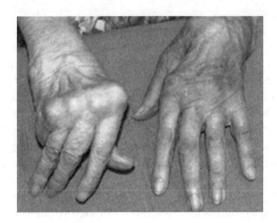

Figure 8.13   Movements after finger joint replacement.
Source: Reproduced with permission from Knight, J., M.D., The Hand and Wrist Institute. "Reconstructive hand surgery". (online) Available at: <http://www .handandwristinstitute.com/reconstructive-hand-surgery-los-angeles/> Date accessed: 12th October 2012. Copyright 2012.

intramedullary-stemmed 'mushroom'-shaped implant. The stem is fitted into the shaft of the first metacarpal (Swanson 1972).

Flexible silicone interposition arthroplasties in the ulnar finger digits do not appear to need revision; they act mainly as spacers rather than joints and often fracture over time, with little clinical significance. Figure 8.13 shows typical movement of the fingers after FJR.

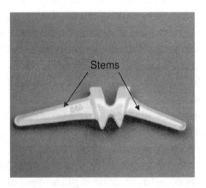

Stems

Figure 8.14   Main features of a silastic finger joint implant.

Figure 8.15   Industrial example of an interphalangeal finger joint replacement. *Source:* Reproduced by permission of Nakashima Medical Co.

Figure 8.14 shows the main parts of a silastic finger joint replacement. It is composed of a monoblock of inert silicone elastomer ('silastic'), which flexes at the junction between its two stems. An industrial example of a finger joint replacement is presented in Figure 8.15.

8.6 Hip

The femoral head and acetabulum are covered by articular cartilage, which allows painless and smooth motion of the joint. When the hip is diseased or injured, the articular cartilage suffers degeneration and wear. The joint surfaces become rough and irregular. Stiffness of the joint and pain ensue. If the pain is severe and all other treatments have been discounted, joint replacement is then considered.

Attempts at hip replacements began in the early twentieth century. Damaged or disfigured joint surfaces were first contoured to facilitate motion. The joint was re-surfaced by an interpositional layer composed of materials such as gold foil or fibrous tissue taken from elsewhere in the body.

These attempts were largely unsuccessful owing to pain and stiffness. In 1923, an alternative technique of 'mould arthroplasty' was introduced.

Moulds made from glass, Pyrex and Bakelite were first tried until eventually Vitallium was adopted owing to its durability. Acetabulum cup arthroplasty became the preferred technique for hip reconstruction.

Heat-cured acrylic femoral head prostheses were another step, but fragmentation, wear, adverse tissue reaction and bone loss proved to be major drawbacks.

Metallic endoprostheses were then introduced, with medullary stems for skeletal fixation. Longer stems enabled transmission of weight-bearing forces along the axis of the femur. Press-fit fixation was used both in the femur and the acetabulum; however cemented implants still result in the best long-term results. Femoral bone loss following loosening was common. Bone erosion on the pelvic side prompted investigations of re-surfacing of the acetabulum.

Many of these drawbacks in THR were most successfully overcome by Charnley in the 1950s and 1960s. His developments are the foundation for present-day THR, although he was not the first, nor the only, developer of implants. This brief account summarizes his approach.

8.6.1 Charnley's Development of Total Hip Replacement (THR)

The problem of boundary lubrication due to friction had to be tackled. To that end, Charnley first lined the acetabulum with a plastic shell and introduced a metallic cup to the femoral head (although he later abandoned this approach, owing to necrosis of the femoral head).

Next, he cemented the stem of a femoral prosthesis and the plastic cup with poly(methyl methacrylate) (PMMA), to secure the components to the bone and allow bone ingrowth for ultimate long term stability of the implant. Additionally, by this procedure stress was more uniformly transferred to a larger bone surface.

He reduced the diameter of the femoral head from the 40 mm size normally used to 22 mm. The effect was to reduce the moment, or the lever arm, of the frictional force, thereby lowering resistance to movement and the shear between the cup and the pelvic bone. Charnley considered that it was preferable to reduce frictional torque and for the cup to have a thicker wall (to allow wear over a longer period before failure), rather than retain a larger head, which would have decreased the pressure per unit surface, and hence lessened linear wear but increased volumetric wear. He adopted high density polyethylene (HDPE) and later ultra-high molecular weight polyethylene (UHMWPE) in place of polytetrafluorethylene (PTFE), a softer material that wore too quickly, in order to counter high wear and adverse tissue reaction.

By 1970, Charnley's pioneering THR yielded vastly improved function with less wear, and greater relief from pain (Charnley 1970). (Nonetheless, more than five years after Charnley's hip replacements were introduced, there were signs of loosening of the fixation of the trochanter, stem failure, and protrusion of the cup.) His efforts stimulated fresh efforts to improve understanding of the biomechanics involved in hip joint replacement. The following section is a summary of the salient biomechanics.

8.6.2 Biomechanics after Hip Joint Replacement

In Chapter 3, the biomechanics of the hip joint have been discussed. In particular, the position of the centre of gravity of the hip in relation to that of the pelvis has been noted to change with movement of the upper part of the body. The forces on the hip joint are applied in the sagittal as well as the coronal plane. Posterior bending of the stem of that part of the joint replacement can arise from the effects of these forces and was seen in early implants made of mild steel.

Movements which flex the hip joint, such as those associated with rising from a chair, mounting or descending stairs or slopes, or keeping the leg straight whilst raising it, all cause increase in the forces acting in the sagittal direction. The stem is then deflected in the posterior direction as noted from Figure 8.16. It should be noted that the forces across the joint are considerable, often many times body weight, because of the muscle forces necessary to effect changes in position or maintain equilibrium.

Simultaneously, the stem is deflected medially by the effect of the forces on the hip, acting in the coronal plane. The combined effect of the sagittal and coronal forces is torsion of the stem. Torsional stability of the femoral part of the hip joint replacement is paramount. Excessive physical movement and high body weight increase the forces on the stem, which may become loose, or bend or even break. The original biomechanical models were rudimentary and did not take into account the muscle forces. Current modelling is much more sophisticated.

The biomechanics of Charnley's THR may now be summarized. The lever arms associated with the body weight and abductors are decreased (medializing the centre of rotation will decrease the distance of the centre of gravity of the body from the fulcrum of the hip). The muscle power is increased by surgery (through deepening of the acetabulum by centralizing the position of the femoral head and lateral reattachment of the greater trochanter, increasing the distance of the muscle attachment from the fulcrum). The ratio of

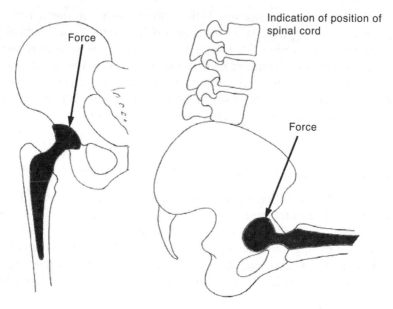

Figure 8.16   Torsional stem forces.
Source: Reproduced from Canale, S.T. and Beaty, J.H. (eds) (2007) *Campbell's Operative Orthopaedics,* 11 edn, Mosby Publishers. Maryland Heights, Missouri.

the two lever arms then tends to unity, and the load on the hip joint can be reduced by as much as 30% due to the surgery.

Both cement-free and cemented joint replacements have been considered. A wider proximal part of the stem will also improve the torsional stability of the femoral component, when cement is not used. Enhanced rotational stability is secured (without the need for cement) when hip-replacement parts are provided with longitudinal cutting flutes or geometrical shapes designed to prevent rotation. (Some failed devices were screwed in and subsequently unscrewed themselves through these rotational forces.) Porous coatings on femoral stems add to these geometrical stability improvements.

Following Charnley's work, a comprehensive biomechanical analysis of total hip replacement was undertaken by Griffiths *et al.* (1971), Swanson (1972), Swanson *et al.* (1973) and Swanson and Freeman (1977).

In one study, they examined a cemented femoral component of a THR. By studying the bending moments they were able to attribute fractured stems to excessive tensile strength of the lateral side, caused by bending. Based on their analysis, new femoral stems with reduced curvature were designed. (More recent models include loading due to muscle force, improving the accuracy of these findings.) Even with progress from these new designs, drawbacks were still found. For instance, although the magnitude of the bending stress in the femoral stem was reduced, the load and stress carried by the cortical bone itself was reduced as the stem was now carrying a higher proportion of the applied load. As bone strength depends on loading, according to Wolff's law, a decrease in its mechanical loading would lead to osteoporosis.

Other components of hip prostheses were also investigated. For example, Swanson *et al.* (1973) reported that the neck of the femoral component of the hip is affected by compression, shear and bending moments. For a specified direction of force on the femoral head, a neck that is nearly horizontal will be affected by larger bending forces and stresses. Yet, a neck, the axis of which runs along the line of action of the resultant force on the femoral head, would be subjected to compressive, but not bending or shear, stresses. It would significantly reduce the abductor lever arm, increasing joint reaction force, and greatly increasing joint surface wear, as was observed in early devices used in difficult cases, where the natural hip had not properly developed. By detailed consideration of the biomechanics, improved designs of hip-joint replacement continued to be developed, as is also evident from the work of English and Dowson (1981); but see also Hamilton *et al.* (2013).

8.6.3 Further Developments in THR Engineering

Engineering improvements in THR continue to be sought. The availability of modern materials has transformed hip replacement procedures. For example, fracture of the femoral stem has been almost eliminated by the use of superalloys. Further reduction in wear of the articulating surfaces has been addressed by investigation of ceramic-to-ceramic, and metal-to-metal articulations, providing low coefficients of friction and reduced wear properties. However, both have introduced fresh issues, such as metal wear debris reactions and ceramic squeaking. On the other hand, articulating surfaces incorporating titanium alloy have proved unsuitable due to rapid wear. Femoral heads made of titanium alloy are no longer used. Increasingly cobalt-chromium alloys are used, owing to their ability to provide a very smooth finish compared with stainless steel.

Metal-on-metal total hip implants, introduced to reduce wear, brought new problems: friction, pain and the development of toxic local levels of

metal ions. Failures of metal-on-metal THR have been discussed by Bolland
et al. (2011) and Smith *et al.* (2012). The former discuss the advantages of
femoral heads of large diameter and hip resurfacing acetabular components
for the active and young (mean age 58 years) suffering from degenerative
joint disease. Increases in cobalt ion levels in the tissue and bloodstream,
due to wear debris, were associated with the need for revision. Pitting and
corrosion was observed on the stem surface. Surface wear rates of 1.55 to
1.86 mm/yr between the head and the cup were found. Revision surgery was
required in more than 8.6% of these cases.

Rieker (2011) suggests that metal-on-metal bearings of smaller diameter
have much less wear (below $1\,mm^3$/year) than the more-established larger
sized bearings. Adverse tissue reactions such as necrosis would then be consequently smaller.

Kazi and Carroll (2011) have reported ceramic-on-metal bearings in THR,
indicating that it could be effective, at least in the short-term.

Loosening has been a common cause for revision of hip-joint replacements.
It was originally associated with cement failure, however biological failure of
fixation can occur with cemented and uncemented prosthesis. The problem
has been tackled for uncemented prostheses by attempts at better press-fit to
improve initial mechanical stability, and porous- and hydroxyapatite-coated
stems and cups to encourage biological ingrowth by creating an attractive
surface and environment. In cemented prostheses better preparation and use
of cement have resulted in more successful outcomes.

Femoral fixation can be improved by injection of cement, occlusion of the
medullary canal, reduction of porosity by pressurization of the cement, and
centralization of the stem. These are all designed to obtain effective conditions
that encourage ingrowth. Such procedures have proved to be less effective
for acetabular fixation.

Aseptic failure of acetabular fixation is a dominant cause in revision of
THR, and can affect 2 to 12% of patients. Late aseptic loosening is largely
induced by wear debris. (Early aseptic loosening is usually associated with
a failure to obtain sufficient stability to allow ingrowth from host bone.) The
debris is mainly generated from the prosthetic joint articular surface, which
can in turn cause sustained chronic inflammatory effects.

The onset of micromotion ranging from $2\,\mu m$ to $15\,\mu m$ in hip replacements with eventual loosening and subsequent reason for revision has been
discussed by Morlock *et al.* (2011). This group investigated modular stems
with additional modularity to the head taper for titanium alloy and cobalt-
chromium alloy neck adaptors, with greater micromotion being encountered
with the former metal. They report that the Australian Arthroplasty Register

(2011) reveals almost twice the risk of revision with modular total hip arthroplasty, due to loosening and lysis.

In hip revision, impaction bone grafting is used to replace lost bone stock as an alternative to milled human bone used as an allograft. High molecular weight polymers have been proposed by Tayton *et al.* (2011) as possessing suitable mechanical properties and cell compatibility. Huntley and Christie (2004) describe procedures for surface preparation in hip joint revision surgery.

The computer-aided methods of manufacture of joint replacements discussed in Chapter 7 have also been used to advance hip joint replacement surgery. For example, Picard *et al.* (2004) and Murphy *et al.* (1986) reported using a three-dimensional reconstruction of the hip of a patient with a congenital hip dislocation, from which design conditions for an implant were selected for subsequent production on a CNC milling machine.

8.6.4 Industrial Examples of THR

Figure 8.17 is a schematic illustration of the main components of a hip joint replacement.

The prosthesis for THR consists of femoral and acetabular components, as shown in Figures 8.18 and 8.19.

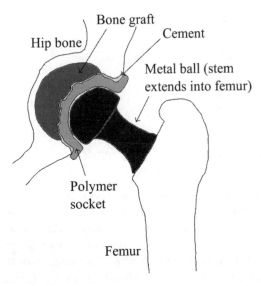

Figure 8.17   Main parts of a hip joint replacement.

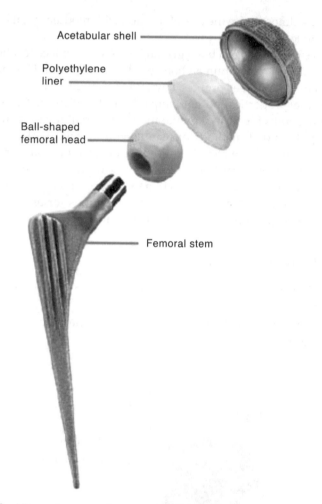

Acetabular shell

Polyethylene
liner

Ball-shaped
femoral head

Femoral stem

Figure 8.18 Hip replacement components.
Source: Reproduced by permission from Zimmer Inc.

The femoral part consists of a head, neck and shaft. The head can be made from cobalt-chromium alloy, alumina, or zirconia. Its shaft may be manufactured from titanium, or cobalt-chromium alloys. (Previously stainless steel type 316L was popular.) The shaft is fixed by cement or press-fit into the medullary canal, which is reamed prior to the insertion of the implant. The acetabular component is made from ultra-high molecular weight polyethylene (UHMWPE).

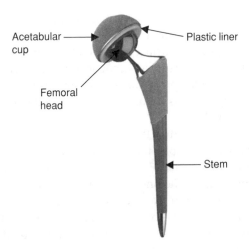

Acetabular cup

Plastic liner

Femoral head

Stem

Figure 8.19 Hip replacement components.
Source: Reproduced with permission from Smith & Nephew, Inc.

8.7 Knee

Pain and stiffness, and associated limitation in the normal arc of motion of the knee in extension and flexion during everyday activity, and even during rest and at night, are common indicators that often lead to the need for total knee replacement (TKR). Of these symptoms, osteoarthritis is the most frequent underlying cause. Age, activity and body weight all contribute to cause and severity.

In some patients, patellofemoral arthritis can become so painful and lack of mobility so progressive that TKR becomes almost the only way to bring relief. Excessive physical activity in the young can impose so much stress on the knee that TKR may become necessary. Trauma can damage the bearing surfaces of the normal knee, which will result in premature wear.

Obesity is another increasingly relevant cause; TKR may not even be possible until body weight is sufficiently reduced. Even if TKR can be performed, heavy body weight can reduce the lifetime of the replacement by increasing the load on the artificial bearing surfaces.

Rheumatoid arthritis is another common factor. Total knee replacement may also be needed owing to other conditions resembling OA and RA that are evident from deformities of the knee.

Arthroplasty to restore the functions of the knee was attempted in the 1860s. Fergusson (1861) resected a knee, partially removing the patella, in order to restore the usefulness of the joint, a so-called resection arthroplasty. This created unacceptable instability.

The restoration of a joint surface was notably difficult. Verneuil (1863) described the use of joint capsule as tissue inserted between raw bone ends after joint resection. Similar procedures involving use of muscle, fat, fascia and man-made fibres, such as nylon, were attempted even until the 1960s, before being abandoned.

Further work on replacement of the entire knee has been described by Lexer (1925), including a knee transplant, undertaken in 1909, from an amputation, for a patient with sarcoma of the joint. He performed knee transplants over the next 20 years, reporting, on the one hand, satisfactory function and absence of pain and, on the other, failure and undesirable complications.

Total knee replacement based on a hinge design spanned the period 1940s to 1960s. An early version consisted of a simple hinge, made of acrylic, and later Vitallium, with a single axle and separate femoral parts. The former was provided with a flange for articulation with the patella. Stability on the bone ends was achieved through inclusion of a straight intramedullary stem (about 100 mm or more in length).

Loosening and stiffness proved problematic. Loosening was reduced by fixation with PMMA bone cement. Torsional stresses about the long axis still proved difficult to absorb, although in the tibia, on the femoral side, the shape of the prosthesis did not create difficulties.

Knee-replacement techniques were advanced further by Parrish (1966) and others using allografts. Alternatives to such methods involving arthroplasty include knee replacement with endoprostheses incorporating re-surfacing of the tibia or femur.

In the 1970s non-constrained knee arthroplasty with cemented prosthetic parts was vigorously pursued (in Wrightington where Charnley worked on the hip) (Gunston (1971)). Two metallic semi-circular runners were inserted into slots in the femoral condyles. These slots articulated into two high density polyethylene tracks that were cemented into slots in the tibial plateau. (Note that the knee replacement did not involve the patellofemoral joint.)

Gunston's designs were refined throughout the 1970s, although the pros-thesis still consisted of a metal femoral component and polyethylene tibial counter surfaces, fixed to the bone by a layer of cement. Fins, grooves and slots on the undersurface of the prosthesis facilitated the tibial fixation of the polyethylene.

A metallic tibial 'tray' that supports the plastic articulating surface of the tibial component of the replacement was introduced in 1974. The tray featured stems on its metallic parts to improve fixation, and held interchangeable polyethylene inserts. With metal components already in place, the polyethylene tibial surfaces enabled more ready adjustments for stability and alignment. Worn polyethylene could be replaced without the cement being affected. A central ball and socket was tried to allow some rotational stability and reduce loosening. However, the complexities of alignment made this configuration difficult to insert and the design was abandoned due to stiffness.

A 'total condylar' design was established in the mid-1970s. The distal femur was replaced by a metallic component, which articulated with sockets in a polyethylene tibial part. A tibial polyethylene peg for fixation was included. Despite further improvements throughout the 1980s, failure still occurred at the bone-cement interface adjacent to the tibia or femur. A plastic polyethylene patellar surface is preferred if the patella is replaced. However, randomized studies suggest that this is rarely necessary (Bourne *et al.* 1995). Porous coatings of the femoral, tibial and patellar implants continue to be investigated. Such coatings are considered to be effective in the fixation of femoral components. They are less so for the tibial tray because initial mechanical stability (necessary to allow bone ingrowth) is more difficult to achieve on the tibia.

Fixation by cement, mostly PMMA, has been a significant improvement. The bone cement transfers loads on the joint from the prosthesis. Two interfaces arise: bone-cement, cement-implant to the bone. The interface between the bone and cement plays a key part.

8.7.1 Biomechanics in Total Knee Replacement

Assessing the effects of arthritis on the biomechanics of the knee is a useful way of understanding how TKR can become necessary.

Normally a load line taken from the centre of the hip to the centre of the ankle runs through the middle of the knee joint. As cartilage is lost from the medial femoral condyle and the tibial plateau due to osteoarthritis of the medial compartment the knee becomes displaced laterally from the line of the centre of gravity of the torso. The partial body weight now acts on a longer moment arm than is usual. As indicated in Figure 8.20, the resultant of the two forces is moved to the medial plateau. W represents the partial body weight held in equilibrium by the lateral tension band T; R is the resultant force. The increased force acts through an area that is smaller than normal. Thus the stress on the medial compartment rises, causing accelerated degeneration of tissue. When arthritis occurs in the lateral compartment, alignment of the lower limb becomes 'valgus' – that is, bent outwards away from the midline.

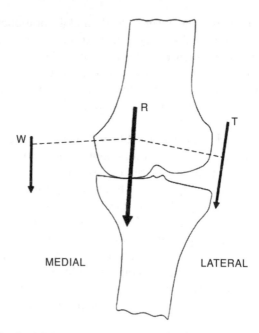

MEDIAL   LATERAL

Figure 8.20 Effect of OA on medial part of knee; increase in resultant force R in medial compartment.
Source: Reproduced with permission from Laskin, R.S. (ed.) (1991) *Total Knee Replacement,* Springer-Verlag Publishers, London.

In order to maintain equilibrium, the resulting load vector moves to the lateral side. The contact stress on the lateral femoral condyle and tibial plateau rises. In consequence, the arthritis condition is exacerbated. Degeneration of the tissue and associated pain, and inhibited mechanical movement of the knee may be overcome through total knee replacement.

The lifetime of a TKR is dependent on the means by which it is fixed to remaining bone. The methods used have to distribute the stress at the bone such that fresh bone will be formed. Excess stress can fatigue the bone, causing loosening. Too low a stress can lead to stress shielding, and bone resorption.

Implants are generally fixed by use of cement, biological ingrowth, or press-fit methods, with the screws and pegs being used to secure the implant.

Most TKR uses cement techniques. Nonetheless effective fixation of tibial components without cement has been claimed. Porous ingrowth components are becoming popular. Their use is dependent on the design of the prosthesis, the precision of its fit, and the alignment of the limb.

The distribution of strain in the proximal tibia is known to be affected by TKR. Metal stems are considered to have a better chance than plastic stems in reducing the strain on the tibia. Micromotion at the fixation site can be reduced by the use of a metal tray. Nonetheless a totally rigid fixation is not needed, as all prostheses have some micromotion. Elimination of loosening of implants depends on the choice of materials, the design of the articular surface, precision instrumentation, and fixation techniques.

The precise insertion of the prosthesis to give the right orientation to the knee is vital for stability. When the knee replacement implant is inserted, there can be as many as six possible positions for each part, of which only one is correct. Thus, when two knee replacements are implanted, there are 36 possible combinations, with only one being correct. This restriction in choice requires that the cuts needed to restore the normal joint line, and re-establish the right tension in the remaining ligaments, as well as preserving the posterior tilt of the tibial component, are performed precisely. Tension in the remaining ligaments, joint capsule and musculature all contribute to the stability. Most investigations of TKR stability involve measurement of axial loading across the implant with shear and torsional loading of one of the components. Until recently this procedure was only carried out with the knee straight. More insight has been gained by testing with the knee flexed.

Gait analysis and video fluoroscopy (radiographs used in real time) are often employed. For example, one study on the use of semiconstrained and nonconstrained implants revealed abnormalities in gait for both devices. Least constrained implants were found to provide the widest range of motion for climbing stairs. In comparison, posterior cruciate resecting prostheses yielded poor results for the same action.

When semiconstrained implants are used, the anterior cruciate ligament is usually removed. The stability of the knee in the anteroposterior plane is then lessened. Without some degree of constraint incorporated into the design of the knee bearing, sliding and rocking can occur in the tibial component in certain positions, such as mid-flexion. The anterior and posterior portions of the implant may be subjected to cyclic tension and compression. Anterior sliding of the tibia on the femur can be countered by employment of a semiconstrained implant within the articular configuration. Nonetheless a disadvantage may then be the transmission of shearing forces to the interface between the implant and bone. A rise in rate of loosening is the consequence. The more an implant is designed to take over the role of soft tissue, the greater the shear at the prosthesis interface. In general, implants perform well under compression but less effectively in tension.

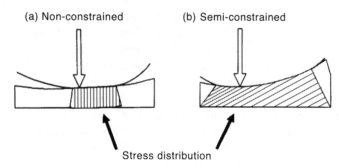

Figure 8.21 Influence on stress of contact area between tibial and femoral components for nonconstrained and- semiconstrained knee devices. Note that the contact area of the nonconstrained knee device is less than the semiconstrained device. Stress is increased through tibial region.
Source: Reproduced with permission from Laskin, R.S. (ed.) (1991) *Total Knee Replacement,* Springer-Verlag Publishers, London.

Upon reducing the contact area between the femoral and tibial components, wear increases in the UHMWPE, reducing the fatigue life of the tibial component owing to point loading and subsurface shear on movement. Mechanical or aseptic loosening can arise. The debris from the wear can lead to inflammatory effects and in turn aseptic loosening. Some tests have shown that some contact stresses in the tibial component can be greater than the compressive strength of the plastic.

Thus stress on the UHMWPE component depends on the area of contact. Stress distributed throughout the tibial component should preferably be not greater than the yield strength of the UHMWPE. For a device with a low contact area, the large compressive forces are transmitted to the tibial surface through a smaller part of the UHMWPE. When the contact area is increased, forces are transmitted through a large surface area, as illustrated in Figure 8.21. However, the wear volume is higher and greater shear forces are transmitted to the bone cement interface. The maximum area of stress within the polyethylene is just below the surface due to deformation on movement.

A uniform distribution of stress at the interface of the prosthesis – cement-bone provides the best conditions for fixation.

As noted from Figure 8.22 (a) and (b), the thicker the tibial component, the lower the stress distributed at the interface. (Implants of thickness less than 6 mm may be too flexible and deform at the interface.) On increase in the thickness of the component, the effective lever arm to shearing forces at the

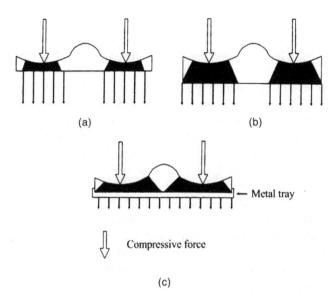

(a)   (b)

— Metal tray

Compressive force

(c)

Figure 8.22   (a) (b) increase in thickness of tibial component distributes stresses through greater volume of polyethylene (c) metal tray imparts stiffer foundation to polyethylene. Stress now distributes more uniformly across bone prosthesis interface.
Source: Reproduced with permission from Laskin, R.S. (ed.) (1991) *Total Knee Replacement,* Springer-Verlag Publishers, London.

articulating surface increases. The moment on the tibial interface increases, and loosening of the component becomes more likely.

Loosening and failure of TKR is mainly due to excessive stress and micromotion at the interface between bone and prosthesis, provided it has been properly inserted.

The stress distribution at the interface of bone and prosthesis plays a key part in determining this lifetime of the TKR. When the stress is greater than the failure strength of the components, their neighbouring structures and that of the prosthesis – bone interface, micromotion of the component will ensue. Ultra-high molecular weight polyethylene is not a rigid material. Thus stress distribution in particular, in the tibial component, is a key element in its loosening and potential failure.

Many experimental and theoretical (mainly finite element) analyses of the stresses arising at the interface of the bone/prostheses have been undertaken. The effects of stability due to cemented and biological in-growth have been compared.

A uniform distribution of stress across the tibial surface can be obtained by use of a metal tray, as shown in Figure 8.22 (c), which distributes the stress uniformly across the distal surface of the tibial component, in contrast to the behaviour obtained with UHMWPE.

The stem of a hinged knee replacement is affected by bending loads, for example when its motion is limited by flexion. Similarly bending loads act on the stem of the knee prosthesis, caused by abducting or adducting moments on the lower leg which are resisted by the rest of the body. The stem has to transmit all these stresses – the former bending loads act in the sagittal plane, the latter in the frontal. They all occur dynamically and are difficult to analyse. Loaded in flexion (stair climbing), the component will be compressed by forces applied to the tibial plateau, through the patella, an intramedullary stem, or spikes fixed by cement in the cancellous bone of the femoral condyles. These forces bend the femoral component, increasing its curvature, and will be transmitted to the femur bone. The femoral component of the prosthesis will undergo compression (for an ideal attachment of femur to component). If the fixation has been defective, parts of the prosthetic component may be unsupported, and bending stresses will arise.

As the need for TKR continues to rise, engineering research and development are proceeding in parallel with clinical developments. For example, Dennis (2009) has evaluated the kinematics of TKR for over 1000 subjects; Walker et al. (2009) draw attention to new designs for TKR, emphasizing the need for restoration of normal knee mechanics; New and Barrett (2009) discuss the effects of cement pressure and penetration on the tibia for TKR; Greenwald, Morra and Rosca (2009) have studied the need for special designs for instance of high knee flexion. The need for revision total knee replacements is regularly monitored, see for example in the Scottish Arthroplasty Project Biennial Report (2012). McLean et al. (2004) also emphasize the effect of the geometry of the revision implant upon the level of micromotion.

8.7.2 Industrial Examples of TKR

Figure 8.23 illustrates the main features of a TKR.

Figure 8.24 shows the main components of knee prostheses. The femoral component can be produced from titanium, cobalt-chromium alloys, or stainless steel. Ceramic surface coatings used on the femoral component include zirconia, alumina, titanium nitride, or diamond-like carbon (DLC). Ultra-high molecular weight polyethylene or cross-linked UHMWPE comprises the polymer insert. The tibial tray can be manufactured from trabecular metal or titanium alloy, or made from polyethylene. Ultra-high molecular weight

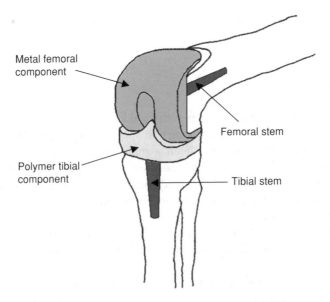

Figure 8.23 Illustration of main parts of a TKR.

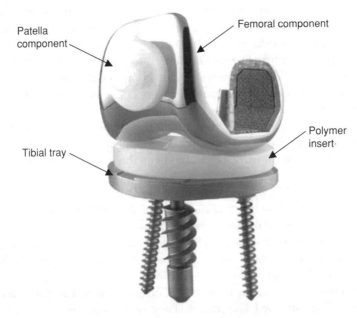

Figure 8.24 Industrial example of a knee joint replacement.
Source: Reproduced by permission of Nakashima Medical Co.

polyethylene, backed with a trabecular metal attachment, is used for the patella replacement.

8.8 Ankle

Arthritis in various forms can affect the ankle. Nonetheless, that resulting from trauma is more common, causing pain, abnormal gait and limited physical movement. Arthrodesis has been a main route to dealing with these problems. Its drawbacks have included failure and need for re-operation, overloading of neighbouring joints and consequential further deterioration. As an alternative, ankle joint replacement may offer improved ankle movement, and better gait.

Vickerstaff *et al.* (2007) have reviewed the progress of total ankle replacements (TAR), as have Gougoulis *et al.* (2008) and Henne and Anderson (2002).

In the 1970s early attempts involved two-part TAR with a polyethylene tibial concave articular surface and usually a cobalt-chromium metallic convex talar component, together with cemented fixation. Constrained (hinge-like motion in one-axis) and unconstrained (multi-axis ball and socket movement) devices were employed. Between 27% and 90% of the constrained TAR prostheses were reported to have loosened after respectively two and ten years. This poor performance was attributed to inadequate cement fixation, unduly high constraints and high stress on components with small areas of fixation. In another type, cylindrical TAR were designed in attempts to recreate the anatomy of the ankle. Unconstrained TAR also loosened due partly to ineffective cement fixation. Other difficulties stemming from the thinness of the soft tissues of the ankle and associated lengthy times for the healing of ankle wounds meant that arthrodesis has been the main way to treat arthritis of this joint.

Kempson *et al.* (1975) investigated ankle implants, applying a compressive vertical load of 1800 N at twisting moments of 40 to 63 Nm to natural ankles, which duly failed. Similar tests were performed on an implanted prosthesis which failed at moments between 29 and 40 Nm. Pain in healthy ankles is known to occur at moments above approximately 22 Nm. In the 1990s, the prospects for TAR were revived.

Ankle replacements are usually based on uncemented fixation, and the avoidance of constrained kinematics. They use porous bead-like coatings with a covering of hydroxyapatite (HA). The HA promotes bone growth upon the implant, therefore the interface between bone and implant should then be more sound.

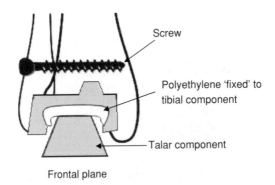

Frontal plane

Figure 8.25 Illustration of Agility fixed-bearing (two-component) ankle joint replacement.
Source: Reproduced from Gougoulias, N.E., Khanna, A., and Maffulli, N. (2008) History and evolution in total ankle arthroplasty. *British Medical Bulletin* **89**, 111–151.

In Figure 8.25 an ankle replacement, the 'Agility' model, is shown. This is based on a two-component, fixed-bearing system. It provides axial rotation, and limited medial-lateral shift. The medial and lateral malleolar surfaces are replaced with this type of ankle replacement. Loosening of the prosthesis is reduced by arthrodesis of the distal tibiofibular syndesmosis on insertion of TAR to eliminate fibular motion. Screws through the distal fibula into the distal tibia ensure more transfer of weight to the fibula (which is about 17% of the total body weight). Bone resection in the longitudinal and transverse dimensions provides the basis of a larger area for bone regrowth.

The alternative to this type of replacement is based on a three-component mobile bearing concept. A flat plate with porous backing acts as the distal tibial articular surface. It articulates with a metal talar dome component through a mobile bearing. The bearing is flat on its tibial surface and conforms on the talar surface. The medial and lateral malleoli are unsurfaced. Bone resection is less common than that for the two-component type. See Figure 8.26.

The ankle joint has a titanium tibial part with a sintered titanium base, the talar section is made from cobalt-chromium alloy with sintered cobalt-chromium beads. A polyethylene insert is locked into the tibial component.

Total ankle replacements for arthritis remain the subject of continuing research and development. Reports of 'Mayo' implants indicate their survival rate was 79% falling to 61% at respectively 5 and 15 years. Revision of failed TAR is becoming more feasible as implant designs continue to be developed (Henne and Anderson 2002), although talar damage is difficult to reconstruct.

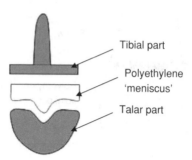

Tibial part

Polyethylene
'meniscus'

Talar part

Figure 8.26 Illustration of three-component ankle joint replacement (note meniscal-type polyethylene is mobile and articulates with both tibial and talar parts).
Source: Reproduced from Gougoulias, N.E., Khanna, A., and Maffulli, N. (2008) History and evolution in total ankle arthroplasty. *British Medical Bulletin* **89**, 111–151.

Harris *et al.* (2011) describe more than 200 ankle replacements over five years, based on a three-component mobile bearing design. Assessment of 140 replacements from 2003 to 2006 yielded a 97% success rate after 20 months. Figure 8.27 shows the position of an ankle joint replacement.

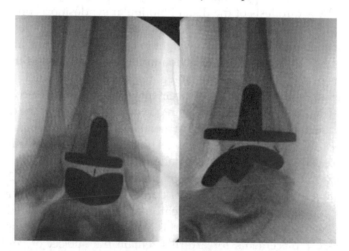

Figure 8.27 X-ray of ankle joint replacement.
Source: Reproduced with permission. Harris, N. (2012). "Chapter 33 – Total Ankle Arthroplasty" in "Practical Procedures in Elective Orthopaedic Surgery, Part V". Springer-Verlag, London. p. 272.

Figure 8.28   Industrial example of an ankle joint replacement.
Source: Reproduced by permission of Nakashima Medical Co. Ltd.

The tibial and talar parts of the ankle shown in Figure 8.28 can be produced from cobalt-chromium or titanium alloys, or stainless steel (all coated with porous metal). The mobile bearing is made from UHMWPE.

8.9 Foot and Toe

Synovitis of the metatarsophalangeal joints caused by RA can eventually lead to destruction of articular cartilage and weakening between ligament and bone. The intrinsic muscles become less able to flex the joint and to extend the middle and distal joints of the digits. The foot becomes deformed. Arthroplasty of the forefoot is aimed at correcting this deformity. It should extend range of movement and relief from pain. Murphy (2003) summarizes the development of arthroplasty of the forefoot from 1951 to 2003. He describes 142 operations performed on 100 patients in 1973 with reports of pain relief in 71% of these cases. In 1984 this figure became 85 to 90%. He indicates the steps being taken to provide a uniform weight-bearing surface, and how Clayton in 1960 made this joint replacement more widely used. In 1980 the metatarsophalangeal joints were replaced with flexible double stem elastomer implants based on the Swanson implants used in the hand and wrist. Further improvements have continued into the 2000s. He concludes that satisfactory joint replacement can be achieved with 80 to 90% of patients treated although, with time and the relatively high loads involved, performance may be impaired and this percentage forecast could be lower.

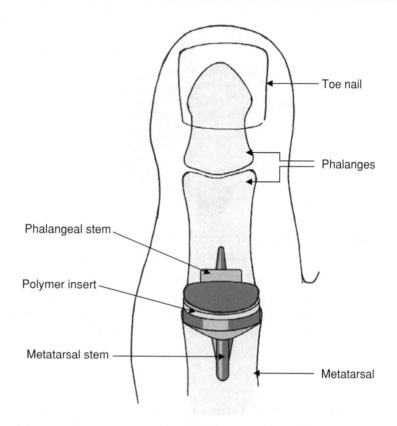

Figure 8.29   Schematic illustration of a (big) toe joint replacement.

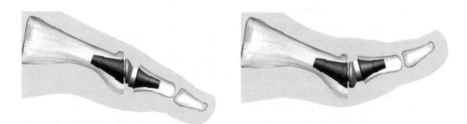

Figure 8.30   Movement of big toe implant.
Source: Reproduced by permission of Medin, A.S.

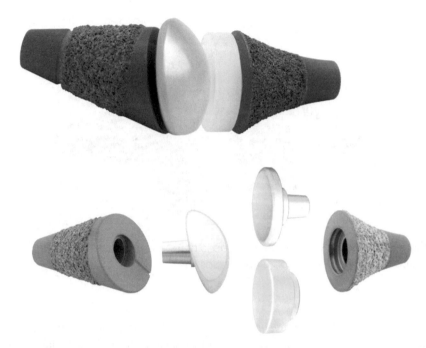

Figure 8.31   Industrial example of a toe joint replacement.
Source: Reproduced by permission of Medin, A.S.

Murphy (2003) discusses the underlying anatomical biomechanics of the toe that cause deformity and instability, and consequently the need for joint replacement, with relief from pain being reported by as many as 85% of patients. The 'extensor digitorum longus' tendon provides the main extensor force on the toe joint. The metatarsophalangeal joint has to be in a neutral or flexed position for this tendon to extend the interphalangeal toe joints. For conditions such as wearing of high-heeled shoes, which extend the toe, the force imposed by the extensor digitorum longus deforms the metatarsophalangeal joint. The latter joint receives its capacity for flexion mainly from the intrinsic muscles. When the joint undergoes chronic extension, the force of the tendons has a deforming effect on the 'dorsal subluxation' of the joint. Stability for the joint stems from the static restraints provided by the collateral ligaments and the plantar plate of the toe. As well as wearing high-heeled shoes, acute hyperextension, such as that encountered by athletes and by excessive weight-bearing, can all lead to instability of the metatarsophalangeal joint, with onset of pain.

Polymer insert

Metatarsal stem

Phalangeal stem

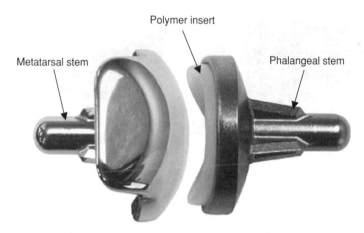

Figure 8.32 Industrial example of a toe joint replacement.
Source: Copyright Ascension Orthopedics, Inc.

When synovitis is also present due to RA or OA, the toe joint can degener-
ate. Its replacement then has to be considered. An example of a replacement
metatarsophalangeal toe joint is shown in Figure 8.29. The movement then
possible with an implant is illustrated in Figure 8.30. Typical components of
a toe joint replacement are presented in Figure 8.31. An example of the entire
assembly is shown in Figure 8.32. The device is inserted without cement. It is
anchored in position by means of a Ti6Al4V titanium part coated with HA.
The head is manufactured from cobalt-chromium-molybdenum alloy, and
the distal elements from UHMWPE.

References

American Academy of Orthopaedic Surgeons (2011) Shoulder Joint Replace-
    ment, http://orthoinfo.aaos.org/topic.cfm?topic=A00094 (accessed 5 May
    2013).
Andersson, G.B.J., Freeman, M.A.R. and Swanson, S.A.V. (1972) Loosening of the
    cemented acetabular cup in total hip replacement. *Journal of Bone and Joint Surgery*
    54B (4), 590–599.
Bolland, B.J.R.F., Culliford, D.J., Langton, D.J. *et al.* (2011) High failure rates with a large
    diameter hybrid metal on metal total hip replacement. Institution of Mechanical
    Engineers (IMechE), event proceedings, Engineers and Surgeons: Joined at the Hip
    III, pp. 37–38.

Bourne, R.B., Rorabeck, C.H., Vaz, M. *et al.* (1995) Resurfacing versus not resurfacing the patella during total knee replacement. *Clinical Orthopaedics and Related Research* **321**, 156–161.

BS EN 980:2008 (2008) *Determination of Endurance Properties and Performance of Stemmed Femoral Components*, British Standards Institution, London.

Canale, S.T. and Beaty, J.H. (eds) (2007) *Campbell's Operative Orthopaedics*, 11th edn, Mosby. Maryland Heights, Missouri.

Charnley, J. (1970) (ed.) Total hip replacement. *Clinical Orthopaedics and Related Research* **72**, 1–241.

Cohen, D. (2012) How a fake hip showed up failings in European device regulation. *British Medical Journal* **345** (7090).

Coonrad, R.W. and Morrey, B.F. (1998) Coonrad/Morrey total elbow: surgical technique, in *Joint Replacement in the Shoulder and Elbow* (ed. W.A. Wallace), Butterworth Heinemann, Oxford.

Dennis, D.A. (2009) Total knee arthroplasty: what I've learned in 15 years. Institution of Mechanical Engineers (IMechE), event proceedings, Knee Arthroplasty 2009: From Early Intervention to Revision, pp. 3–4.

English, T.A. and Dowson, D. (1981) Total hip replacement, in *Introduction to the Biomechanics of Joints and Joint Replacement* (eds D. Dowson and V. Wright), Mechanical Engineering Publications Ltd., London.

Everitt, H., Elliott, M., Bigsby, R. *et al.* (2011) Comparison between friction and lubrication behaviour of large diameter ZTA ceramic vs CFR-PEEK and MOM hip resurfacing. Institution of Mechanical Engineers (IMechE), event proceedings, Engineers and Surgeons: Joined at the Hip III, p. 79.

Ewald, F.C. (1975) Total elbow replacement. *Orthopedic Clinics of North America* **6** (3). 685–696.

Fergusson, W. (1861) Excision of knee joint: recovery with a false joint and a useful limb. *Medical Times and Gazette* **1**, 601.

Gougoulias, N.E., Khanna, A. and Maffulli, N. (2008) History and evolution in total ankle arthroplasty. *British Medical Bulletin* **89**, 111–151.

Greenwald, S., Morra, E. and Rosca, M. (2009) Kinematic performance of high flexion knee designs. Institution of Mechanical Engineers (IMechE), event proceedings, Knee Arthroplasty 2009: From Early Intervention to Revision, pp. 37–38.

Griffiths, W.E.G., Swanson, S.A.V. and Freeman, M.A.R. (1971) Experimental fatigue fracture of the human cadaveric femoral neck. *Journal of Bone and Joint Surgery* **53B** (1), pp. 136–143.

Gunston, F.H. (1971) Polycentric knee arthroplasty. *Journal of Bone and Joint Surgery* **53**, 272–275.

Hamilton, D.F., Simpson, A.H, Howie, C.R. *et al.* (2013) The role of the surgeon in the application of the scientific method to new orthopaedic devices. *Surgery*, **11** (3), 117–119.

Harris, N.J., Sturdee, S.W. and Farndon, M. (2011) *Total Ankle Replacement. The Early Results of 140 Consecutive Cases of the AES Prosthesis*, Department of Orthopaedics and Trauma, The General Infirmary at Leeds.

Henne, T.D. and Anderson, J.G. (2002) Total ankle arthroplasty – a historical perspective. *Foot and Ankle Clinics of North America* **7**, 695–702.

Hunsicker, P. (1955) Arm Strength at Selected Degrees of Elbow Flexion, Technical Report, 45-548. Wright Patterson Air Force Base. Dayton, Ohio.

Huntley, J.S. and Christie, J. (2004) Surface replacement of the hip: a late revision. *Canadian Journal of Surgery* **47** (4), 302–303.

Inui, A., Kokubu, T., Fujioka, H. *et al.* (2011) Application of layered poly (L-lactic acid) cell free scaffold in a rabbit rotator cuff defect model. *Sports Medicine, Arthroscopy, Rehabilitation, Therapy and Technology* **3** (29). DOI: 10.1186/1758-2555-3-29.

Jennings, L.M., Al-Hajjar, M., Begand, S. *et al.* (2011) Wear of novel ceramic-on-ceramic bearings under adverse and clinically relevant hip simulator conditions. Institution of Mechanical Engineers (IMechE), event proceedings, Engineers and Surgeons: Joined at the Hip III, pp. 81–83.

Kazi, H.A. and Carroll, F.A. (2011) Use of a ceramic on metal bearing in total hip arthroplasty. Institution of Mechanical Engineers (IMechE), event proceedings, Engineers and Surgeons: Joined at the Hip III, pp. 39–43.

Kempson, G.E., Freeman, M.A. and Tuke, M.A. (1975) Engineering considerations in the design of an ankle joint. *Journal of Biomedical Engineering* **10** (5), 166–171.

Kudo, H. and Iwano, K. (1990) Total elbow arthroplasty with a non-constrained surface-replacement prosthesis in patients who have rheumatoid arthritis. A long-term follow-up study. *Journal of Bone and Joint Surgery* **72-A**, 355–362.

Laskin, R.S. (ed.) (1991) *Total Knee Replacement*, Springer-Verlag, London.

Lexer, E. (1925) Joint transplantation and arthroplasty. *Journal of Surgery, Gynaecology and Obstetrics* **40**, 782–809.

McLean, A.J., McGeough, J.A., Simpson, H. and Howie, C.R. (2006) Effects of central stem length on the initial micromotion experienced by the tibial tray in revision total knee arthroplasty. *Journal of Bone and Joint Surgery* **88-B**, 397–398.

Medicines and Healthcare Products Regulatory Agency (2010) Medical Device Alert. Ref: MDA/2010/069. Date issued: 7 September 2010.

Meuli, H.C. (1984). 'Meuli total wrist arthroplasty'. Clinical Orthopaedics and Related Research, Volume 187, pp. 107–111, in Canale, S.T., and Beaty, J.H. (Eds.) (2007) *'Campbell's Operative Orthopaedics'*. 11th Edition, Mosby Publishers (Affiliate of Elsevier).

Morlock, M.M., Jauch, S. and Huber, G. (2011) Modular stems in primary THA – blessing or curse? Institution of Mechanical Engineers (IMechE), event proceedings, Engineers and Surgeons: Joined at the Hip III, pp. 29–32.

Murphy, G.A., (2003) Lesser toe abnormalities, in *Campbell's Operative Orthopaedics*, 10th edn (ed. S.T. Canale), Mosby. Maryland Heights, Missouri. pp. 4047–4083.

Murphy, S.B., Kijewski, P.K., Simon, S.R. *et al.* (1986) Computer-aided simulation, analysis and design in orthopaedic surgery. *Orthopedic Clinics of North America* **17** (4), 637–649.

New, A.M.R. and Barrett, D. (2009) Effects of Surgical Technique on Cement Pressurisation in the Tibia in Total Knee Replacement Institution of Mechanical Engineers

(IMechE), event proceedings, Knee Arthroplasty 2009: From Early Intervention to Revision, pp. 31–35.

NHS (National Health Service) National Services Scotland. Scottish Arthroplasty Project – Biennial Report (2012). ISD (Information Services Division) Scotland Publications, NHS National Services Scotland, Edinburgh.

Parrish, F.F. (1966) Treatment of bone tumours by total excision and replacement with massive autologous and homologous grafts. *Journal of Bone and Joint Surgery* **48**, 969–990.

Picard, F., Moody, J.E., DiGioia, A.M. *et al.* (2004) History of computer-assisted orthopaedic surgery of hip and knee, in *Computer and Robotic Assisted Knee and Hip Surgery* (eds A.M. DiGioia, B. Jaramaz, F. Picard and L.P. Nolte), Oxford University Press, Oxford, pp. 1–22.

Rieker, C. (2011) Why are small-diameter metal-on-metal bearings working fine? Institution of Mechanical Engineers (IMechE), event proceedings, Engineers and Surgeons: Joined at the Hip III, pp. 47–49.

Smith, A.J., Dieppe, P., Vernon, K., Porter, M., Blom, A.W. (2012) "Failure rates of stemmed metal-on-metal hip replacements: analysis of data from the National Joint Registry of England and Wales". *Lancet 2012* **379**, 1199–1204.

Souter, W.A. (1977) 'Total replacement of the arthoplasty of the elbow' in 'Joint Replacement of the Upper Limb', pp. 99–106, Trans. Inst. of Mech. Eng., London.

Souter, W.A., Nicol, A.C. and Paul, J.P. (1985) Anatomical trochlea stirrup arthoplasty of the rheumatoid elbow. *Journal of Bone and Joint Surgery* **67B**, 676.

Swanson, A.B. (1968) Silicone rubber implants for replacement of arthritic or destroyed joints in the hand. *Surgical Clinics of North America* **48**, 1113–1127.

Swanson, A.B. (1972) Disabling arthritis at the base of the thumb: treatment by resection of the trapezium and flexible (silicone) implant arthroplasty. *Journal of Bone and Joint Surgery, American Volume* **54** (3), 456–471.

Swanson, S.A.V. and Freeman, M.A.R. (eds) (1977) *The Scientific Basis of Joint Replacement*, Pitman Medical Publishing Co. Ltd, London.

Swanson, S.A.V., Freeman, M.A.R. and Heath, J.C. (1973) Laboratory tests on total joint replacement prostheses. *Journal of Bone and Joint Surgery* **55B** (4), 759–773.

Tayton, E.R., Fahmy, S., Aarvold, A. *et al.* (2011) An analysis of six polymers for both the mechanical and biocompatibility characteristics required for use as an osteogenic alternative to allograft in impaction bone grafting. Institution of Mechanical Engineers (IMechE), event proceedings, Engineers and Surgeons: Joined at the Hip III, p. 87.

Verneuil, A.S. (1863) Affection articulaire du genou. *Archives of Medical Research*.

Vickerstaff, J.A., Miles, A.W. and Cunningham, J.L. (2007) A brief history of total ankle replacement and a review of current status. *Medical Engineering and Physics* **29**, 1056–1064.

Volz, R.G. (2007) The development of a total wrist arthroplasty, in *Campbell's Operative Orthopaedics* (eds S.T. Canale and J.H. Beaty), 11 edn, Mosby Publishers, vol. 116, pp. 209–214. Maryland Heights, Missouri, U.S.A.

Walker, P., Yildrim, G., Arno, S. and Heller, Y. (2009) New concepts in TKR design. Institution of Mechanical Engineers (IMechE), event proceedings, Knee Arthroplasty 2009: From Early Intervention to Revision, p. 25.

Wallace, W.A. (ed.) (1998) *Joint Replacement in the Shoulder and Elbow*, Butterworth Heinemann, Oxford.

Wilkinson, J.M. (2012) Metal-on-metal hip prostheses: where are we now? *British Medical Journal* **345** (7885), p. 10.

Wright, P.E. (2007) Arthritic hand, in *Campbell's Operative Orthopaedics* (eds S.T. Canale and J.H. Beaty), 11th edn, Mosby Publishers, Maryland Heights, Missouri, U.S.A. pp. 3689–3738.

Index

The Engineering of Human Joint Replacements, First Edition. J.A. McGeough.
© 2013 John Wiley & Sons, Ltd. Published 2013 by John Wiley & Sons, Ltd.